Basics of Electric Motors

Including Polyphase Induction and Synchronous Motors

Second Edition

Anthony J. Pansini, E.E., P.E.
Life Fellow IEEE; Member ASTM

*Engineering and Management Consultant
Waco, Texas*

1421 S. Sheridan
Tulsa, OK 74112

Pennwell Publishing Company
Tulsa, Oklahoma

Library of Congress Cataloging-in-Publication Data

PANSINI, ANTHONY J.

Basics of electric motors: including polyphase induction and synchronous motors/Anthony J. Pansini.

p. cm.

Includes index.

ISBN 0–13–060070–9

1. Electric motors. I. Title.

TK2514.P36 1989

621.46′2—dc19 88–15153
CIP

1421 South Sheridan
Tulsa, Oklahoma 74112

Printed in the United States of America

1 2 3 4 00 99 98 97 96

ISBN 0-87814-673 -3

Dedication

In Memory of Galileo Ferraris, originator of the Rotating Field Principle and the inventor of the induction motor – on the centennial of his death

Contents

Preface

This second edition, essentially a second printing of the original edition, provides an opportunity to include discussion of the effect of harmonics not only on motors, but also on the supply circuits to which they are connected. Harmonics are created by non-resistance loads, including not only motors, but gaseous type lighting, periodic switched equipment, and electronically operated or electronically controlled devices. Proliferation of this type of loading requires that the effect of the wave distortion they cause, distortions that are reflected in the harmonic frequencies imposed on the fundamental alternating current wave, be the subject of the discussion.

Electric motors have made possible industrial exploits that were not considered feasible with human, animal, and steam power. Yet their operation, the variety of types, their applications, and associated problems have remained somewhat of a mystery to those outside the engineering fraternity.

That the basis of motors is the phenomenon of magnetism, familiar to anyone who has at some time toyed with a horseshoe magnet, may indeed be surprising. The simple concept of the action of a motor is that of one set of magnets chasing another set attached to a shaft, causing it to rotate and perform work.

In this text we have deliberately avoided highly technical explanations and procedures, and have made only limited use of mathematics. The material presented is divided into two parts. One part deals exclusively with motors, their types and characteristics, and their operation and maintenance. Another part delves into other properties of electricity and magnetism associated with motor action. The many illustrations help in the understanding of this important and interesting subject.

Mechanics, electricians, technicians, sales and maintenance personnel, and others engaged in dealing with motors will find this work helpful. Others whose interest may also be pertinent will find it instructive. Earlier success with similar presentations concerning power delivery systems and equipment provided the motivation and experience needed to pursue this undertaking.

Grateful acknowledgment is made to the many contributors of drawings, diagrams, and other descriptive material, including extracts from Essentials of Electricity by this author and reproduced by the courtesy of the Long Island Lighting company. Thanks too, are due to many of my friends and associates for their generous aid and encouragement and not least to my beloved wife for her patience and support.

Waco, Texas *Anthony J. Pansini*

PART A Basics of Electric Motors

chapter 1

Electromagnetic Basis

INTRODUCTION

Electric motors convert electrical energy into mechanical work, employing the same phenomenon of magnetism that is exhibited by the small horseshoe magnet, a novelty or toy familiar to almost everyone. The rotation of an elementary motor, producing the mechanical work, is caused by the interaction of the magnetic fields of a fixed magnet, called a *stator*, and a movable magnet, called a *rotor* (or sometimes an *armature*). At least one of the magnetic fields involved is produced by the electrical energy input. The repulsion action between the magnetic fields of the stator and rotor cause the rotor to revolve; the revolving rotor is harnessed to perform mechanical work. (A similar action takes place in electric generators, except that mechanical work is applied as input and electrical energy produced as output.) It is well, therefore, that the phenomena concerning magnetism and electromagnetism be understood.

MAGNETISM

Some substances exhibit a power to attract materials such as iron, steel, nickel, cobalt, and some alloys made of those materials; these substances are known as *magnets* or *magnetic materials*. If suspended so that they may swing or rotate

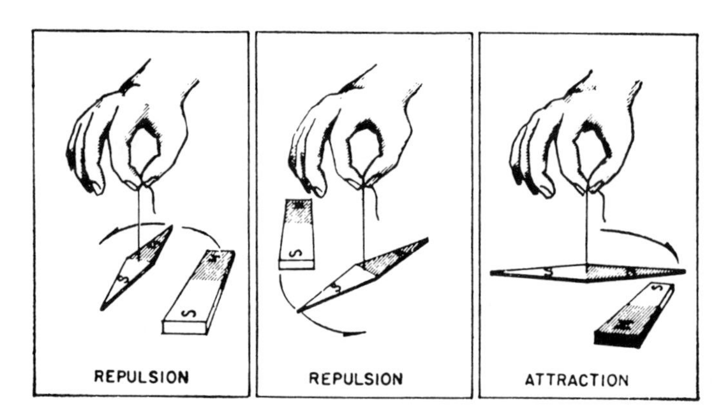

Figure 1–1 Laws of attraction and repulsion. (Courtesy U.S. Navy.)

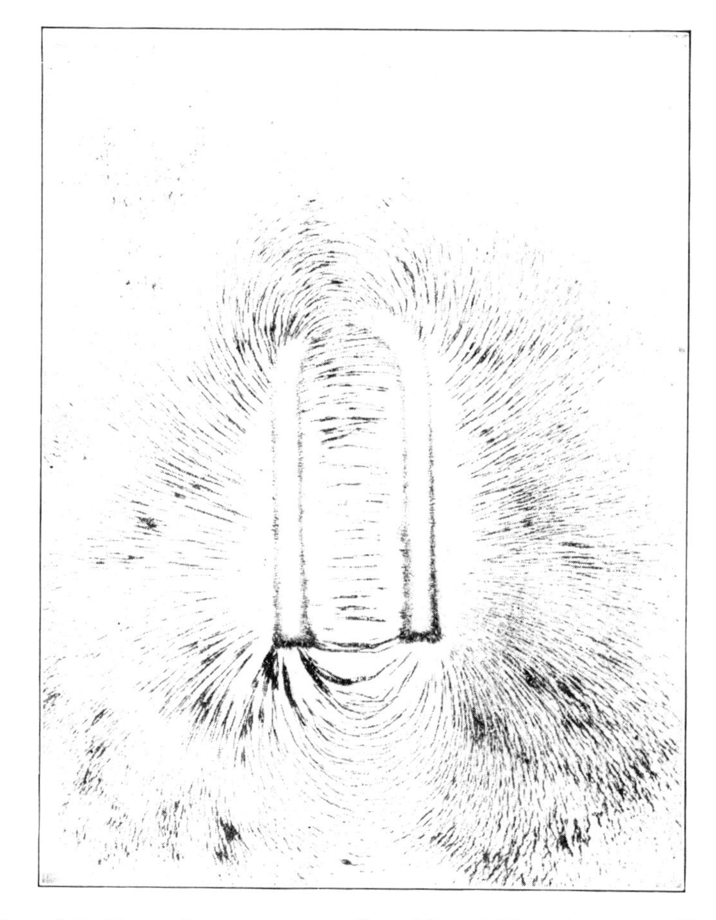

Figure 1–2 Magnet shapes and magnetic lines of force: (a) horseshoe magnet; (b) bar magnet.

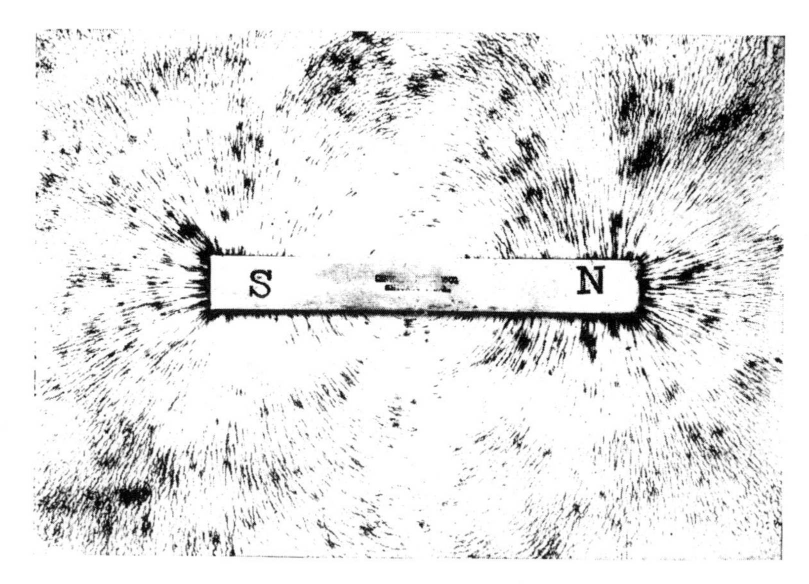

Figure 1–2 ***(Continued)***

freely, pieces of such substances will come to rest in an approximately north–south line of direction. The same end will always point in one direction; that pointing toward the geographic north is referred to as the *north pole* of the magnet, and the other end the *south pole* (Figure 1–1).

The property of attraction can be illustrated by the horseshoe magnet mentioned earlier. Looking at Figure 1–2a, a magnetic field consisting of imaginary (invisible) lines of force may be theorized. For convenience, these lines may be said to emanate from the north pole and enter the south pole, then proceed through the magnet back to the north pole; the path of these lines of force constitutes a complete magnetic circuit. The same phenomenon exists in a bar shaped magnet and in magnets of other shapes (Figure 1–2b).

The phenomenon of magnetism is not well understood. Looking at Figure 1–3, one theory assumes that each of the molecules of some substances are tiny

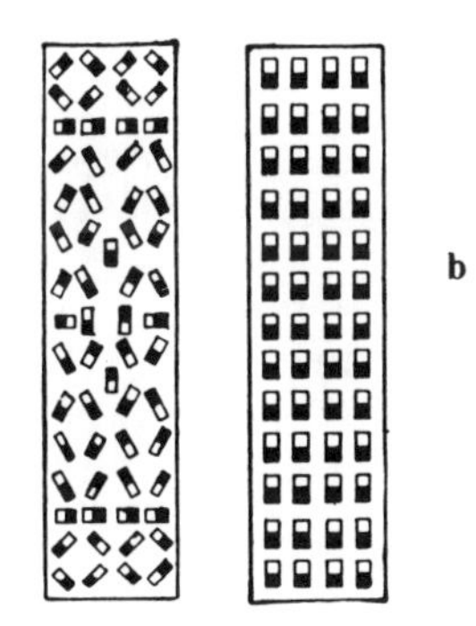

Figure 1–3 Molecular theory of magnetism: (a) unmagnetized iron; (b) magnetized iron.

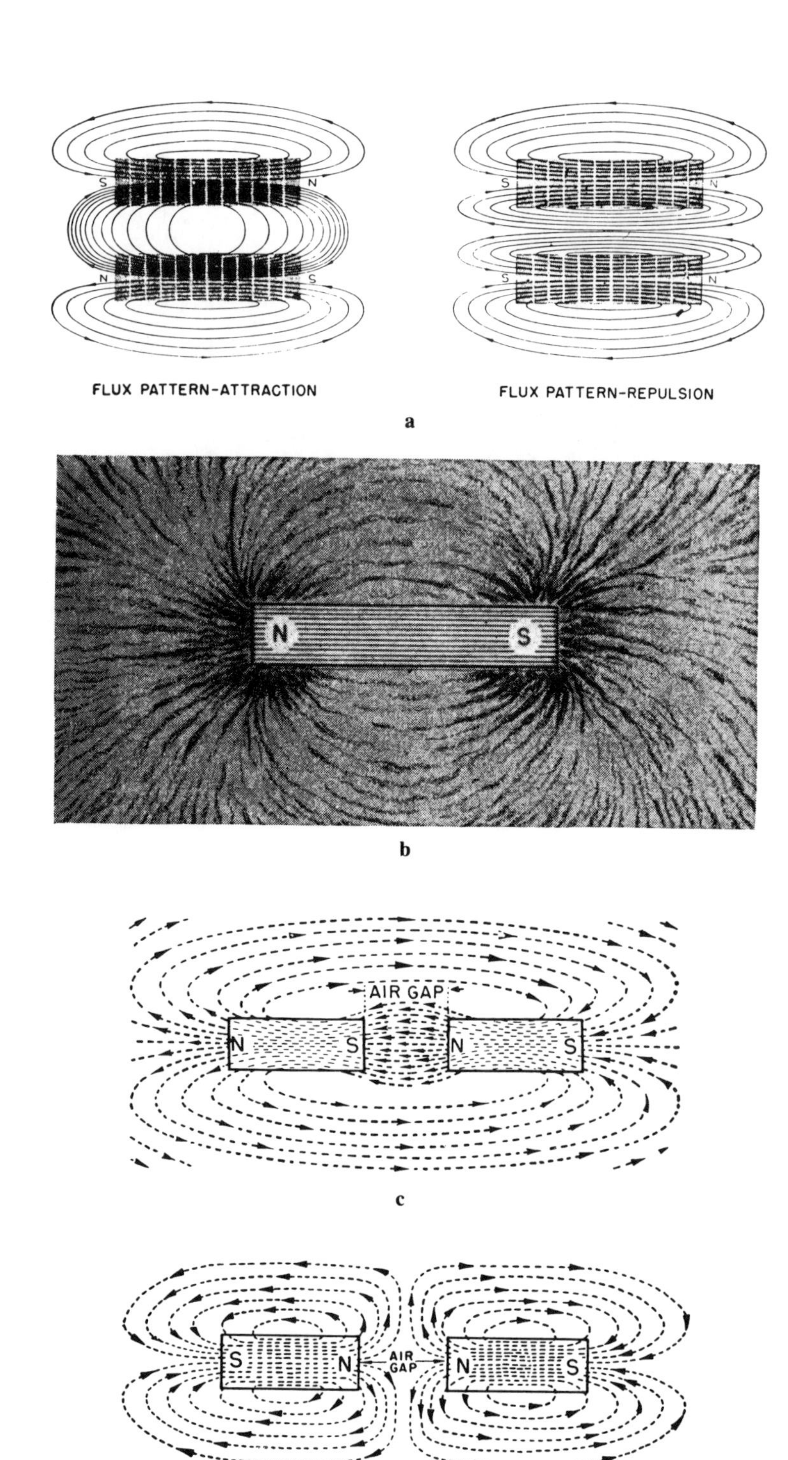

Figure 1–4 Properties of magnetic lines of force: (a) flux patterns of adjacent parallel bar magnets; (b) magnetic field around a magnet; lines of force do not appear to cross each other. (c) Unlike poles attract; (d) like poles repel. (Courtesy U.S. Navy.)

magnets which, when arranged at random, produce no magnetic effect, and when arranged so that the north poles of each of the tiny molecular magnets point in the same direction, produce an external magnetic field or magnetic lines of force (Figure 1–3).

From observation of the behavior of magnets by sprinkling iron filings around a magnet, several facts can be deduced:

1. The magnetic field around a magnet can be considered symmetrical but can be disturbed by the magnetic field of a nearby magnet (Figure 1–4a).
2. The lines of force appear to emanate from one end, or north pole, and enter the other end, or south pole, with the greatest intensity found near the pole surfaces and diminish with increased distance from the poles; the magnetic lines of force appear not to cross each other at any point (Figure 1–4b).
3. Opposite or unlike poles tend to attract each other; like poles tend to repel each other (Figure 1–4c and d). The force of attraction or repulsion varies directly as the product of the separate pole strengths, and inversely as the square of the distance separating the poles. *For example*: If either pole strength is *doubled*, the distance between poles remaining the *same*, the force between the poles will be *doubled*. If the distance between two north poles is *doubled*, the force of repulsion is decreased to *one-fourth* of the original value.

ELECTROMAGNETISM

In 1819, Hans Christian Oersted, a Danish physicist, found that a definite relationship exists between magnetism and electricity. He discovered that an electric current is accompanied by certain magnetic effects.

A current of electricity flowing through a wire produces not only heat but also a magnetic field about the wire; this may be proved by placing a compass needle in the vicinity of the current-carrying wire (Figure 1–5).

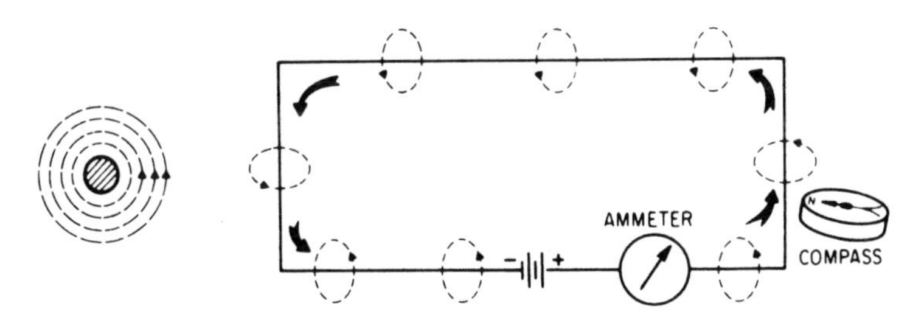

Figure 1–5 Magnet field around a current-carrying conductor.

Figure 1–6 Right-hand-thumb rule.

If the direction of the electric current is assumed to be from negative to positive, it will be observed that the magnetic needle placed adjacent to the conductor will always point with its "north" pole in a certain definite direction. The needle is forced into this position by the magnetic lines of force, sometimes also referred to as the *magnetic field*, and sometimes as *magnetic flux*.

Right-Hand-Thumb Rule

This observation leads to a general rule known as the right-hand-thumb rule. If the wire is grasped in the right hand with the thumb outstretched pointing to the direction of the electric current, the fingers curled around the wire will indicate the direction of the magnetic lines of force (Figure 1–6).

FIELD ABOUT A COIL

A magnetic field around a single wire carrying a current is rather weak. By winding the wire into a ring, the magnetic lines are concentrated in the small space inside the ring or coil and the magnetic effect is much increased. The lines of force are grouped, resulting in a magnetic field stronger than that around the single wire (Figure 1–7a).

A coil of wire may be considered as a succession of these rings stacked one after the other. Each adds its quota to the magnetic field. Most of the magnetic lines of force pass straight through the coil. Each line makes a complete circuit, returning by a path outside the coil. A coil carrying a current is in fact a magnet. Where the lines come out is the "north" pole; where they enter is the "south" pole (Figure 1–7b).

The right-hand-thumb rule may also be applied in determining the polarity of a coil. If the coil is grasped in the right hand with the fingers pointing in the direction of the current flowing around the coil, the outstretched thumb points toward the north pole of the coil (Figure 1–6).

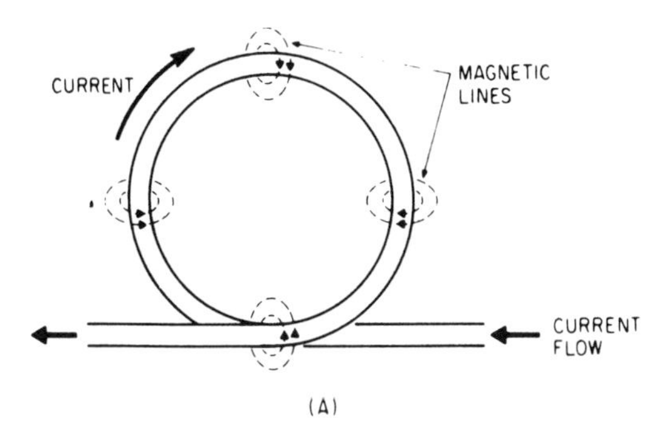

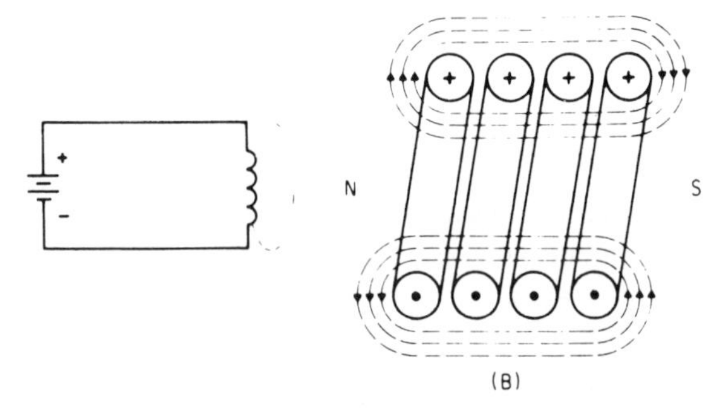

Figure 1–7 (a) Magnetic field about a loop; (b) magnetic field about a coil. (Courtesy Long Island Lighting Co.)

Strength of the Magnetic Field

The strength of the magnetic field inside a coil depends on the strength of the current flowing and the number of turns. It is therefore expressed in *ampere-turns*, that is, amperes multiplied by the number of turns. Thus a single turn carrying a very large current may produce the same effect as a great many turns carrying a small current.

A coil with an air core, however, produces a comparatively weak magnetic field. Its strength can be increased enormously by inserting an iron or steel core inside the coil. This arrangement is generally referred to as an *electromagnet* (or sometimes, *solenoid*).

Example:

Assume a coil of wire of 100 turns having an air core produces 10,000 lines of force when a current of 5 amperes flow through it, that is, by a coil magnetic strength of 5 ampere-turns. If a magnetic field of 20,000 lines of force is desired, it may be achieved by

1. Doubling the number of turns in the coil, keeping the current flowing through it constant:

$$200 \text{ turns} \times 5 \text{ amperes} = 1000 \text{ ampere-turns}$$

2. Keeping the number of turns in the coil the same and doubling the current flowing through it:

$$100 \text{ turns} \times 10 \text{ amperes} = 1000 \text{ ampere-turns}$$

Permeability

When an iron core is used in an electromagnet, it produces a stronger magnetic field than when no core is used; further, if a steel core is used, the magnetic field produced is even stronger than that produced when iron is used

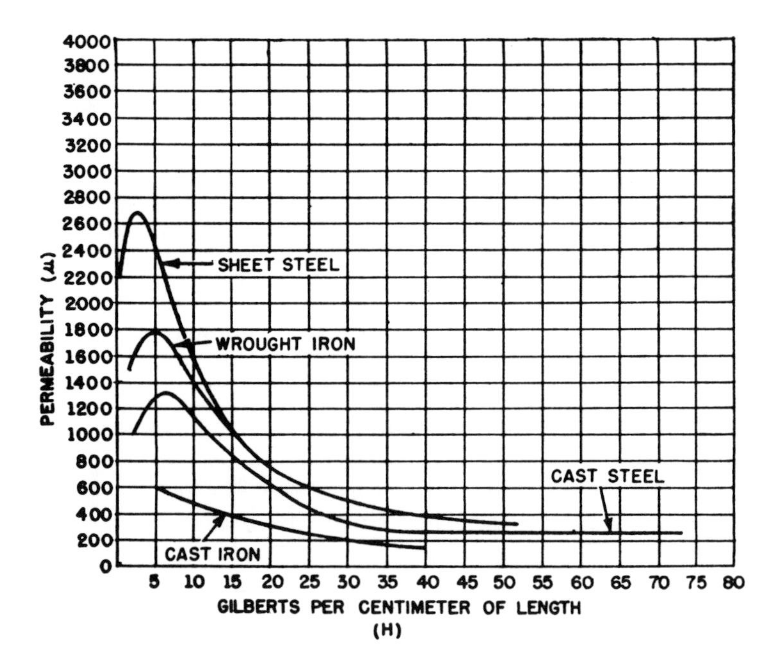

Figure 1–8 Permeability curves. (Courtesy Westinghouse Electric Co.)

as a core. The ratio of the magnetic field, or *flux*, produced by a coil when the core is iron, steel, or some other substance, to the flux produced when the core is air is called the *permeability* of that substance. The premeability of a substance is thus a measure of the relative ability to conduct magnetic lines of force, or its magnetic conductivity.

The permeability of air is taken as 1 or unity, and is essentially the same for nonmagnetic materials such as wood, aluminum, copper, and brass. The permeability of magnetic materials may range from values as low as 200 for cast iron to over 3000 for special steels (Figure 1–8).

Example:

If the coil considered above of 100 turns and 5-ampere current flow, producing 10,000 magnetic lines of force, has inserted in it a cast-iron core whose permeability is 200, the magnetic lines of force produced will be

$$10{,}000 \times 200 = 2{,}000{,}000 \text{ lines}$$

If a steel core having a permeability of 1000 is inserted in the coil, the magnetic field produced will be

$$10{,}000 \times 1000 = 10{,}000{,}000 \text{ lines}$$

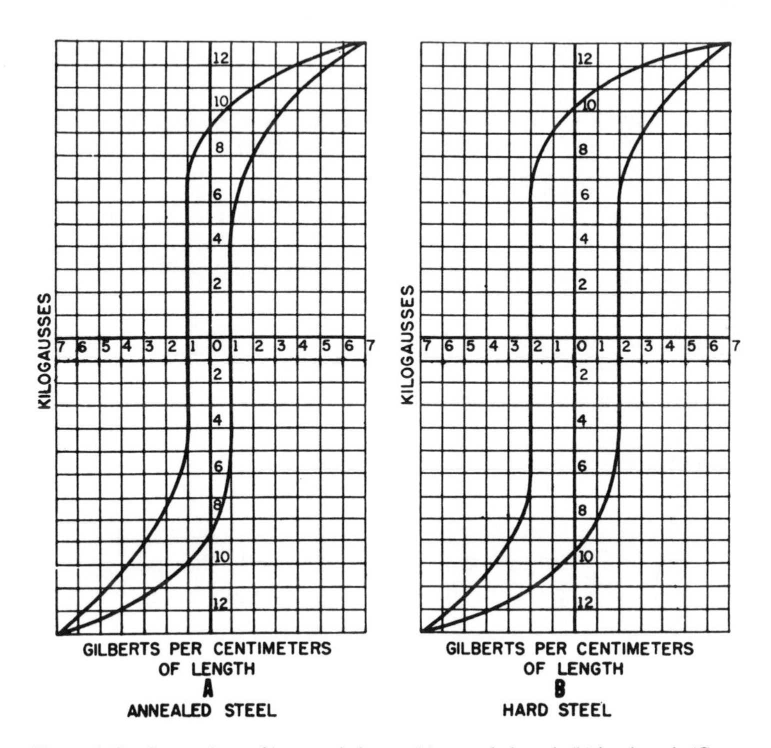

Figure 1–9 Comparison of hysteresis loops: (a) annealed steel; (b) hard steel. (Courtesy Westinghouse Electric Co.)

Hysteresis

Another characteristic of the magnetic field set up by an electromagnet when it is energized and deenergized is the ability of a magnetic substance to retain its magnetism after the magnetizing force has been removed; this is sometimes referred to as the *retentivity* of the magnetic substance. A graphical representation of the magnetization produced by a magnetizing force from a starting point of zero through a maximum in one direction and a maximum in the opposite direction is shown in Figure 1–9. The property of a magnetic substance that causes the magnetization to lag behind the force that produces it is known as *hysteresis*. It is caused by the friction of the molecules as they change their positions within the magnetic material undergoing magnetization.

ELECTROMAGNETIC INDUCTION

In addition to the relation between electricity and magnetism in which magnetism is produced with the aid of electric current, there is another most important relationship: the production of electricity with the aid of magnetism. Although observed by many others, it was not until 1831 that Michael Faraday reported the electromagnetic induction phenomenon in which electricity could be produced.

When a conductor is moved through a magnetic field, an electrical pressure, or voltage, is produced in the conductor, and an electric current will flow if the conductor is part of an electric circuit. The current will, in turn, produce a magnetic field about the conductor which will tend to be repelled by the magnetic field through which the conductor is passing. Hence work will be required to push the conductor through the magnetic field that now acts as a force to resist the movement of the conductor. The energy used in pushing the conductor through the magnetic field is equal (less losses) to the electrical energy generated, or "induced," in the conductor. This phenomenon provides a means of converting mechanical work directly into electricity.

Generator Action

This phenomenon is illustrated in Figure 1–10, which shows a magnetic field established in the air gap between the north and south poles of a magnet and a conductor extending through this field. If the conductor is moved downward between the two poles of the magnet, an electrical pressure, or voltage, is set up in it which attempts to circulate an electric current flowing in the direction indicated, but is prevented from doing so because the circuit is not complete. If the circuit is completed into a loop (as shown by the dashed line), an electric current would flow in the loop. The current would flow only when the wire moved through the magnetic field. The movement of the wire would be resisted

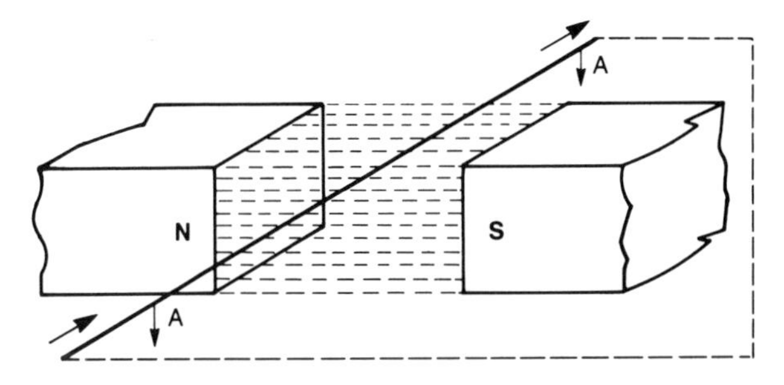

Figure 1–10 Conductor moving in direction A-A through magnetic field.

by the magnetic lines of force just as though the two poles of the magnet were connected by a number of rubber bands.

Factors Affecting Induced Voltage

If the simple magnet is substituted by a more powerful electromagnet, it will be found that a greater electrical pressure, or voltage, is now induced in the conductor cutting the magnetic lines of force. That is, the greater the number of lines of force (or the stronger the magnetic field), the greater will be the voltage induced. More mechanical work, however, will now be required to move the conductor through the stronger magnetic field.

The greater the length of the conductor, the more voltage will be produced because more lines of force will be cut. If the conductor cuts the magnetic lines of force at an angle, it will cut only the same number of lines of force as when it is cutting them at right angles. In considering conductor length, therefore, only the *effective* length, that is, the length that cuts the lines of force at right angles, should be taken into account.

Similarly, if the speed at which the conductor is moved through the magnetic field is increased, it will be found that the voltage induced in the conductor will be also increased. Here, too, more mechanical work will be required to move the conductor at the greater speed through the magnetic field.

The magnitude of the electrical pressure induced in a conductor while it is moving through a magnetic field will therefore be determined by the rate of cutting of the lines of force of the magnetic field. The rate of cutting of the lines of force depends on three factors:

1. The effective length of the conductor that cuts through the magnetic field
2. The speed at which it is moving
3. The strength or density of the magnetic field, that is, the number of lines of force per unit area (square inch or square centimeter)

Example:

Voltage induced in a conductor (E) depends on the strength of the magnetic field (ϕ), the effective length (L) of the conductor, and the speed (S) at which the conductor is cut by the magnetic field; that is,

$$\text{voltage} = \text{field} \times \text{effective length} \times \text{speed}$$

$$E = \emptyset \times L \times S$$

Assume

$\emptyset$ = 50,000 lines per square inch (in.^{2})

L = 10 inches,

S = 3800 inches per second (20 in. diameter, 3600 rpm)

E = 1 volt (V) produced by 100 million lines per square foot per second

Then

$$\text{voltage } (E) = \frac{50{,}000 \times 10 \times 3800}{100{,}000{,}000/144} = 2736 \text{ V}$$

The cutting of magnetic lines of force by a conductor moving through a magnetic field may be compared to the cutting of blades of grass on a lawn. Three factors that determine the rate of cutting of the blades of grass are:

1. The length of the cutting blade of the mower at right angles to the swath cut
2. The speed at which the cutting blade moves through the lawn
3. The density of the lawn; that is, the number of blades of grass per square foot of lawn

A change in any of these three factors makes a corresponding change in the rate of cutting of the blades of grass as well as in the amount of work required to cut the grass. If a two foot cutting blade is used, the rate of cutting as well as the work required will be one-third more than if a one and a half foot cutting blade were used. Here it will be noted that less grass will be cut if the mower is pushed so that the cutting blade is not at right angles to the swath cut. If the speed at which the mower is moved through the lawn is doubled, the rate of cutting is doubled, as is the work necessary to push the mower twice as fast. If the mower is moved through a lawn in which there are many blades in each square foot of lawn, the rate of cutting and the work required are correspondingly greater than in a less dense lawn in which there are fewer blades of grass in each square foot of area.

Relative Motion of Conductor and Field

An electrical pressure, or voltage, can be induced in a conductor by moving it through a magnetic field as described earlier. A voltage can also be induced electromagnetically by moving a magnetic field across the conductor. It does not make any difference whether the conductor is moved across the magnetic field or the magnetic field is moved across the conductor. A stationary conductor that has a magnetic field sweeping across it is cutting the magnetic field just the same as though the conductor were moving across the magnetic field.

Referring to Figure 1–10, it will be observed that the conductor moving downward through the magnetic lines of force whose direction is assumed to be from the north pole to the south pole of the magnet has a voltage induced

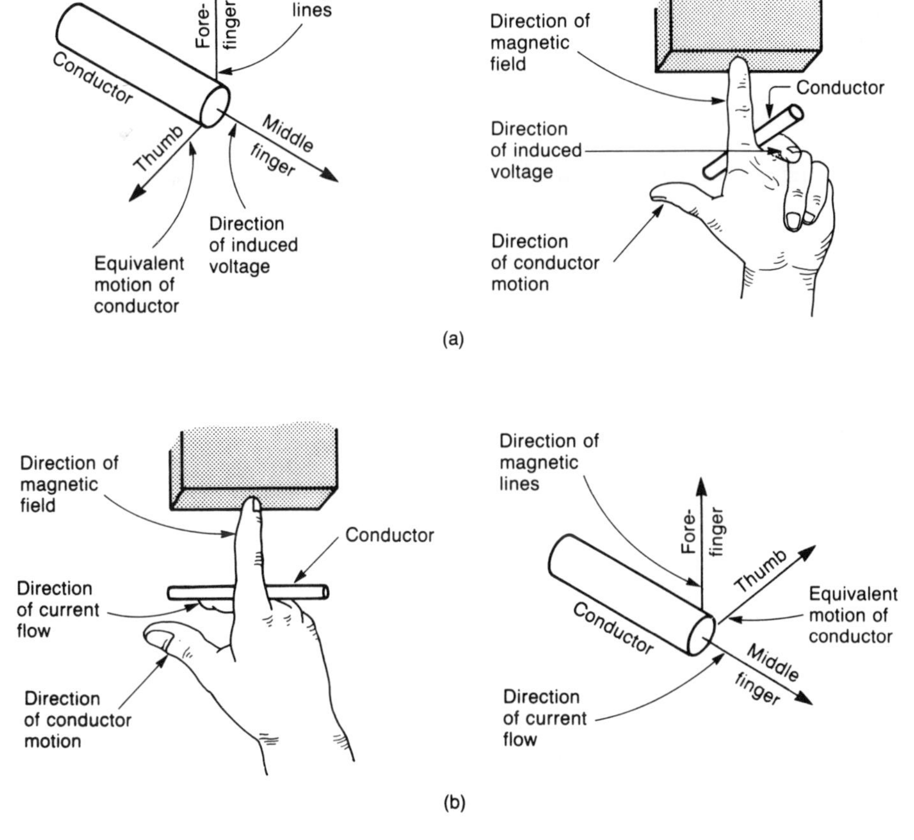

Figure 1–11 (a) Left-hand rule for motor action; (b) Right-hand rule for generator action. (Courtesy Long Island Lighting Co.)

in it such that the electric current is in the direction indicated. If the conductor were moved upward through the same magnetic lines of force, the voltage induced in it will be in the opposite direction to that indicated above. Thus it will be seen that the direction of the induced voltage depends on the direction of the magnetic field and the direction of motion of the conductor.

Right-Hand Rule for Generators

The relations in direction between the magnetic field, the motion of the conductor, and the voltage induced in it may be expressed by the so-called right-hand rule. If the right hand is held so that the thumb, forefinger, and middle finger are all at right angles with each other, as shown in Figure 1–11b, the thumb will indicate the direction of motion of the conductor through the magnetic field, the forefinger will indicate the direction of the magnetic lines of force (issuing from the north pole), and the middle finger will indicate the direction in which a voltage is being generated in the conductor. The middle finger will also indicate the direction in which the current will flow in the conductor.

MOTOR ACTION

Force Acting on a Conductor

Conversely to the generator action described, electrical energy may be converted into mechanical work, involving the same principles. If a conductor carrying an electric current is placed in a magnetic field, there will be a force produced tending to move the conductor. This moving conductor can be harnessed to do some mechanical work.

Left-Hand Rule for Motors

The same general relations that exist between the magnetic field, motion of the conductor, and the electrical pressure (or current) in the conductor, also apply to the latter phenomenon. This is known as *motor action* to distinguish it from the first phenomenon, known as *generator action*. The relationship between the direction of the voltage (or current) applied, the magnetic field, and the motion of the conductor may be expressed by a corresponding left-hand rule, as shown in Figure 1–11a; the quantities represented by the fingers are the same as for the generator right-hand rule, described above.

INTERACTION OF MAGNETIC FIELDS

Rotation of an elementary motor is generally accomplished by the interaction of two magnetic fields: one associated with a stationary magnet and a second associated with a magnet capable of being moved. The stationary field may be established between the poles of a magnet, which may be those of a U-shaped magnet described previously. The movable magnet may be established around a conductor that will carry an electric current; the conductor may be made movable by shaping it in the form of a closed loop and mounting it between fixed pivots.

Figure 1–12a shows the uniform distribution of the main field flux between the two poles of the stationary magnet, and no current flows in the conductor loop. Figure 1–12b shows the magnetic fields surrounding the two conductors constituting the loop when current flows in the direction indicated, without reference to the magnetic field between the poles of the stationary magnet. Figure 1–12c shows the resultant magnetic field produced by the interaction of the two magnetic fields. Note that the flux is strengthened below the conductor

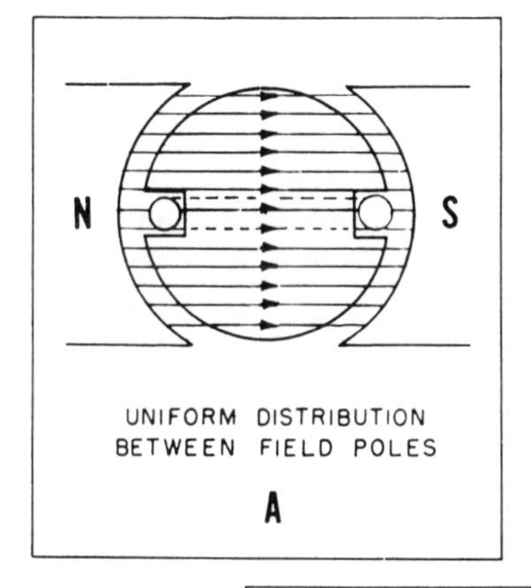

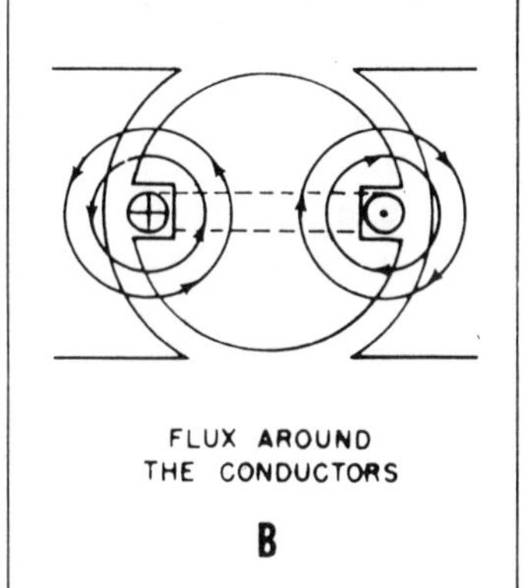

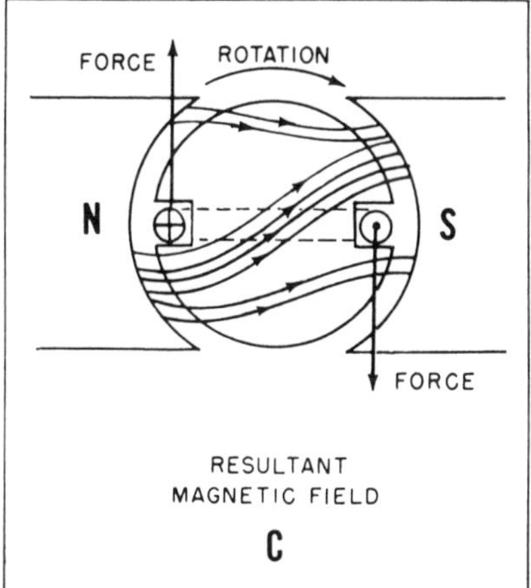

Figure 1–12 Magnetic fields in a motor: (a) uniform distribution between field poles; (b) flux around the conductor; (c) resultant magnetic field. (Courtesy U.S. Navy.)

at the north pole and above the conductor at the south pole because the lines of force are in the same direction at these points. Conversely, the flux is weakened above the conductor at the north pole and below the conductor at the south pole because the lines of force are opposite in direction and tend to cancel each other. The lines of force are like stretched rubber bands and tend to contract, with the result that the loop rotates in a clockwise direction; if the direction of the current in the loop is reversed, the direction the loop will rotate will be reversed.

Factors Affecting Mechanical Work Output

As the output of the elementary motor described above depends on the attractive and repulsive forces between two magnetic fields, the force resulting from the rotation of the loop will be greater the greater the strength of the magnetic fields involved. The strength of the stationary magnetic field may be increased by replacing the magnet with a stronger one. The strength of the magnetic field around the conductors of the loop may be increased by increasing the value of the current flowing in the conductors. The strength of the magnetic field of the loop may also be increased by adding additional conductors to the loop. The interaction of the two strengthened fields, besides resulting in increased mechanical work output, may also manifest itself in an increase in the speed of rotation of the loop. The similarity with the factors affecting the induced voltage in a generator are to be noted.

Torque

Torque may be defined as the tendency of a force to produce rotation about an axis. Applied to motors, it is the turning or twisting effect of the rotor. Mathematically, it is the product of the force acting at the surface of the rotor multiplied by the perpendicular distance to the line of action of the force from the center of rotation of the rotor (usually, the radius of the rotor); Figure 1–12c).

Example:

If the force acting on the current carrying conductor is one pound and the radius of the rotor is 1.5 ft, the torque exerted by each conductor is

$$1 \text{ lb} \times 1.5 \text{ foot} = 1.5 \text{ pound-feet}$$

The total torque of a motor is the sum of the individual torques contributed by all the conductors; if there are 200 active conductors each developing a force of one pound, the total torque will be

$$200 \times 1 \text{ lb} \times 1.5 \text{ foot} = 300 \text{ pound-feet}$$

Horsepower

The horsepower of a motor may be determined by the speed of the rotor in revolutions per minute (or second), the effective rotor radius at which the force acts, and the total force acting at and tangent to this effective radius.

Work is accomplished when force acts through distance. For example, when a force of one pound acts through a distance of one foot, one foot-pound of work is accomplished. By definition, 33,000 foot-pounds of work done in one minute constitutes one horsepower.

Example:

If the motor described above makes 100 revolutions per minute and the effective radius is 1.5 feet, the distance through which the force moves in one revolution of the rotor is

$$\text{circumference} = 2\pi \times \text{radius}$$
$$= 2 \times 3.14 \times 1.5 = 9.42 \text{ feet}$$

Further,

$$\text{work} = \text{force} \times \text{distance}$$
$$= 200 \text{ lb} \times 9.42 \text{ ft} = 1884 \text{ ft-lb (per minute)}$$

For 100 rpm,

$$1884 \times 100 = 188{,}400 \text{ ft-lb}$$

or

$$\frac{188,400 \text{ ft-lb/min}}{33{,}000} = 5.71 \text{ horsepower}$$

COUNTER ELECTROMOTIVE FORCE

Every motor is also a generator. Referring to the action of a generator, as the conductor is moved through the magnetic field, it cuts lines of flux and will have a voltage induced in it (Figure 1–13). Applying the right-hand generator rule, it will be found that the generated voltage is in opposition to the impressed voltage. This counter voltage, or *counter EMF*, is induced in the windings of any rotating electric machinery and always opposes the impressed voltage. This counter EMF is directly proportional to the speed of the rotor and the strength of the magnetic field. The effective voltage in the motor coils is equal to the impressed voltage minus the counter EMF. This phenomenon of self-induction was observed and explained by Joseph Henry in the 1830s when working on electromagnets.

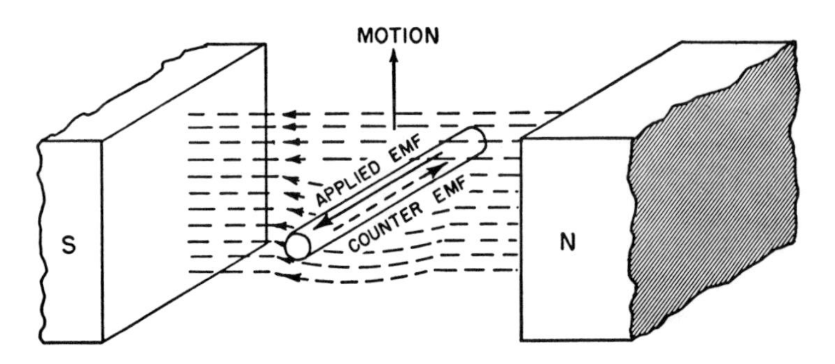

Figure 1–13 Generator action in a motor. (Courtesy U.S. Navy.)

REVIEW

- Some substances, such as iron, steel, nickel, cobalt, and some alloys are known as magnets and, when suspended freely, will tend to come to rest in an approximately north–south direction (Figure 1–1).
- Such substances produce a magnetic field about them that exhibit certain characteristics (Figure 1–4).
    1. The magnetic field around a magnet can be considered symmetrical but can be disturbed by another adjacent magnetic field.
    2. The lines of force appear to emanate from one end, or north pole, and enter at the other end, or south pole, with the greatest intensity found near the pole surfaces; the magnetic lines of force appear not to cross each other at any point.
    3. Opposite poles tend to attract each other, and like poles tend to repel each other. The attractive or repulsive force varies directly as the product of the separate pole strengths, and inversely as the square of the distance separating the poles.
- An electric current flowing through a wire produces a magnetic field about the wire. If the wire is grasped in the right hand with the thumb outstretched pointing to the direction of the electric current, the fingers curled around the wire will indicate the direction of the magnetic lines of force; right-hand rule (Figure 1–6).
- The strength of a magnetic field within a coil depends on the strength of the current flowing and the number of turns; it is expressed in ampere-turns.
- The ability of a material to conduct magnetic lines of force is known as its permeability. The property of a substance that causes its magnetism to lag behind the force that produces it is known as hysteresis (Figure 1–8).

- When a conductor is moved through a magnetic field, an electric pressure or voltage is produced in it and its strength will depend on the strength of the magnetic field, the length of the conductor, and the speed with which it cuts it (Figure 1–10).

- The relations in direction between the magnetic field, the motion of the conductor, and the voltage induced in it are expressed in the right-hand rule (for generators): If the right hand is held so that the thumb, forefinger, and middle finger are at right angles with each other, the thumb indicates the direction of motion of the conductor, the forefinger the direction of the magnetic field, and the middle finger the direction of the voltage generated (Figure 1–11a).

- For motor action, the same relationships can be expressed by the left-hand rule (Figure 1–11b); the quantities expressed by the fingers are the same as for the generator right-hand rule.

- The interaction of two magnetic fields, one produced by a stationary magnet and the other by a movable one, results in a force that tends to have the conductors producing the movable field rotate (Figure 1–12).

- The force tending to make the rotor of a motor revolve, multiplied by its perpendicular distance to the line of action of the force from the center of rotation (usually the radius of the rotor), is known as the torque of the motor and is expressed in pound-feet.

- Work is accomplished when a force acts through a distance; it is expressed in foot-pounds per unit of time. One horsepower is equal to 33,000 foot-pounds per minute.

- Every motor is also a generator. As the rotor turns, its magnetic flux will cut the windings of the field in the stator, producing a voltage in it which will oppose the impressed voltage; this opposing voltage is known as the counter electromotive force, or counter EMF (Figure 1–13).

STUDY QUESTIONS

1. What two effects are produced when a current of electricity flows through a wire?
2. What is the right-hand-thumb rule for determining the direction of the magnetic lines about a conductor?
3. Why is the field inside a ring or a coil of wire stronger than that given by a single straight wire? How may this field be further strengthened?

4. On what two quantities does the strength of the field inside a coil without iron depend?
5. How is the strength of a magnetic field expressed?
6. How may a voltage be induced in a conductor?
7. On what three factors does the magnitude of the induced voltage depend?
8. What is the rule for determining the direction of the voltage induced in a conductor?
9. In what manner does a coil of wire carrying a current of electricity move when placed in a magnetic field?
10. How may the direction a motor will rotate be determined?

chapter 2

Direct-Current Motors

INTRODUCTION

The work of Oersted, Faraday, and Henry provided the necessary basis for the development of the generation of electricity by mechanical means. In summary, these included:

1. A current flowing in a conductor produces around itself a magnetic field whose strength depends on the magnitude of the current.
2. A conductor cut by a magnetic field induces a voltage in that conductor which, in a completed circuit, will cause a current to flow.
3. The voltage induced in a conductor cut by a magnetic field depends on the strength of the magnetic field, the effective length of the conductor, and the speed at which the conductor is cut.
4. The directions of each of these three factors are mutually at right angles to each other, in accordance to the right-hand rule.

From these, afterward, electric motors were to be developed that converted electricity into mechanical work. But it remained for others, Del Negro and Pixii in 1832, to construct practical machines that necessitated the solution of many engineering problems.

MOTOR ACTION

Motor action is developed by the interaction of two magnetic fields that continually oppose each other, arranged so that one of the magnetic fields is capable of revolving. In analyzing and describing the interactions that take place, the bases listed above also apply to motors, with the exception that the factors in the fourth item accord to a left-hand rule. The four bases should be kept continually in mind.

A stationary electromagnet, with several layers of turns, furnishes a strong

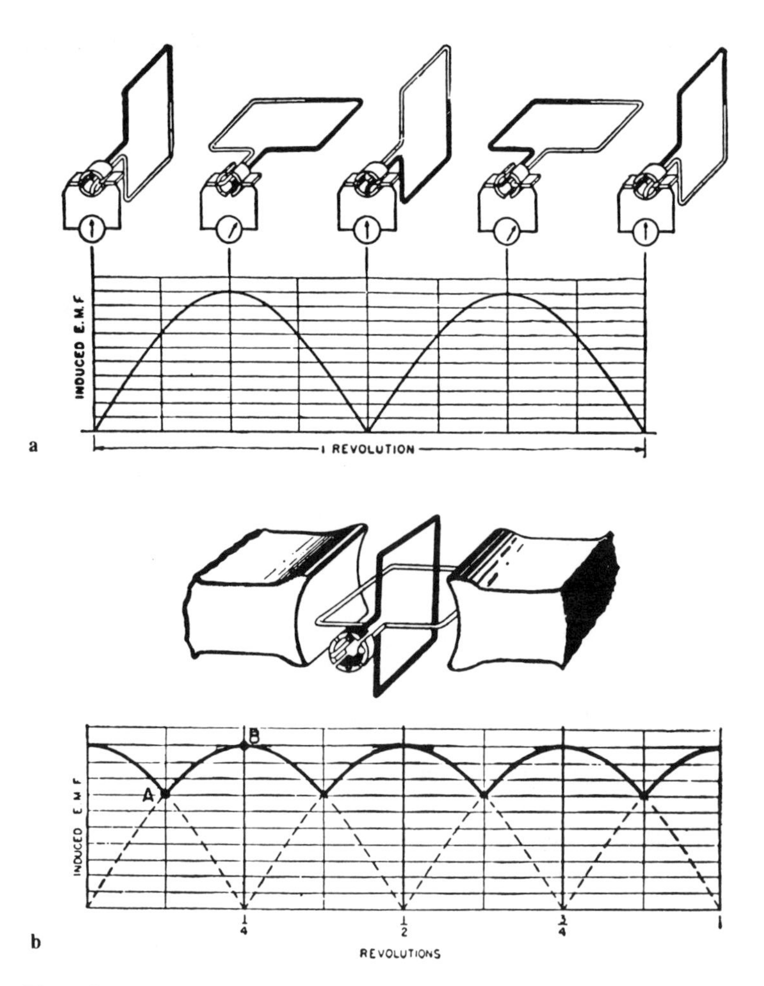

Figure 2–1 Commutator action on armature showing voltage from (a) a single-coil and (b) a two-coil armature. (Courtesy U.S. Navy.)

magnetic field. This magnetic field is further strengthened by a number of such electromagnets. These comprise the "poles" constituting the "field" of the motor. The strength of the rotating field is increased by having the coils of the rotor or armature wound on a steel core and by increasing the number of conductors along the surface of the rotor. Further, the resistance to the "flow" of the magnetic fields is reduced, limiting it generally to the air gap between the stationary and rotating electromagnets.

The strengths of each of the two magnetic fields can be controlled by varying the electric current flowing through the coils producing them; hence the interaction between them can also be controlled. This, in turn, controls the mechanical work output of the motor. The torque, or turning tendency, of an electric motor therefore depends on the strength of the stationary field and that of the magnetic field produced by the current in the armature. Several ways are described further on.

Commutation

The direction of the magnetic field of each of the coils wound on the armature (rotor) will depend on the direction of the flow of current within it. As the flow on one side will be opposite to that on the other side of the coil, the magnetic fields set up will also be in opposite directions. For a constant interaction to take place, producing a force always in one direction permitting the rotor to revolve, it is necessary to change the direction of flow of current in each coil of the rotor during each half turn; the relation between the polarities of the interacting fields can thus be maintained. This is accomplished by the commutator (Figure 2–1). The ripple effect on the voltage is reduced when two coils are used instead of one, as shown in Figure 2–1b. By adding more armature coils, the ripple effect can be further reduced.

ARMATURE

Armatures may be of two types: one known as the Gramme ring (after Zenoble Gramme, who originated it in 1871) and the other, the slotted ring or drum (developed by Antonio Pacinotti in 1860). Although the Gramme armature was a later development, it has been abandoned in favor of the drum type, but some motors still exist that have a Gramme ring armature.

Gramme Ring Type

The Gramme ring armature has a winding of insulated wire wrapped around a steel ring, with taps at regular intervals to the commutator segments (Figure 2–2a); it can be adapted to motors having any number of poles. The active parts of the winding lie on the outer surface of the ring, while those on the inside

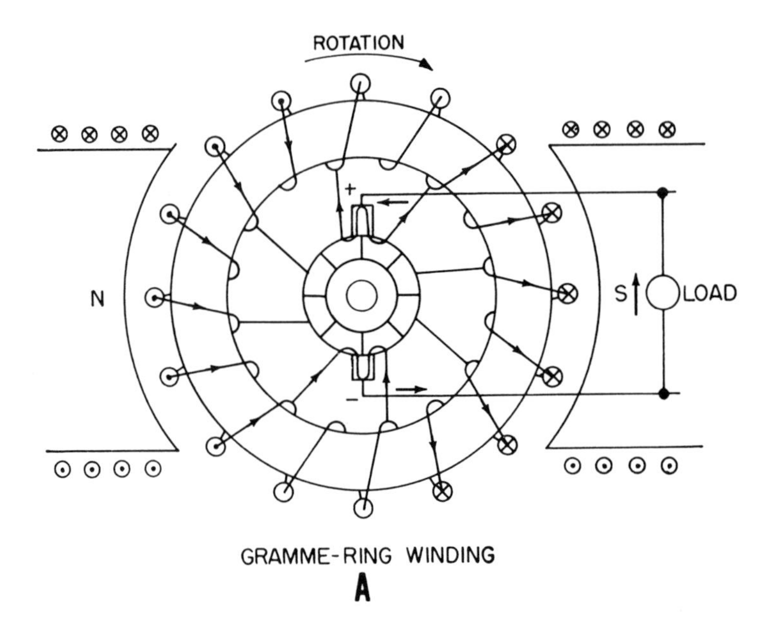

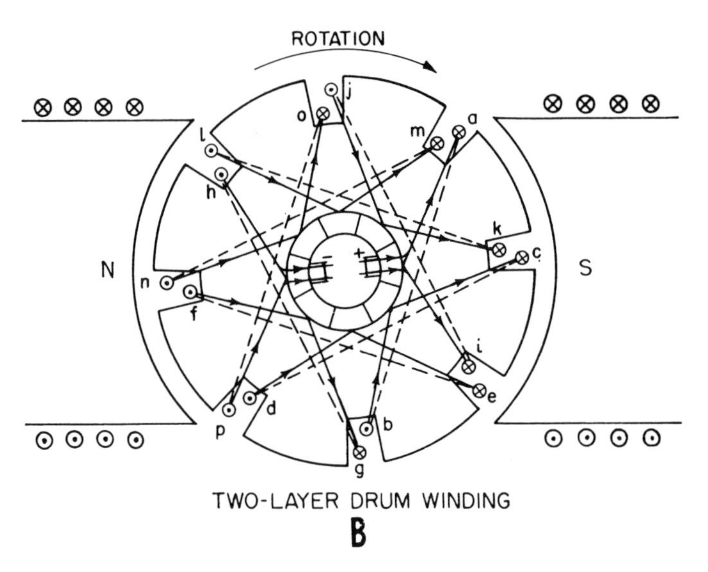

Figure 2–2 Basic direct-current motor armature windings: (a) Gramme ring; (b) two-layer drum. (Courtesy U.S. Navy.)

are cut by practically none of the magnetic flux and act as connectors for the active portions of the winding; it is to be noted that the greater part of the winding is inactive. A schematic diagram is shown in the figure; it is often used as a simplified diagram for any type of armature.

In the diagram for a two-pole machine (Figure 2–2a), there are two parallel paths for the armature current between the brushes. No circulating current flows between the two paths because equal but opposite voltages exist in the two halves of the winding. In this type of armature, the brush axis is perpendicular to the field axis.

Drum Type

The drum armature has the conductors of the winding embedded in slots on the surface, as shown in Figure 2–2b; connections on the back side of the armature between the two halves of each coil are indicated by the dashed lines. The end connections on the front between the coils and commutator segments are indicated by the solid lines. The greater use of conductor material in this type of winding principally accounts for its preference (over the Gramme ring).

Although the brushes are shown inside the commutator surface, they normally make contact on the outside surface. With the exception of the coil end connections, all of the conductor (winding) is active. The same relations exist between the two paths of the windings between the brushes; the voltages are equal and in opposition, and no circulating current flows. In this type of armature, the brush axis coincides with the field axis.

Windings

Direct-current motor armatures are generally wound with preformed coils (Figure 2–3). The distance between one side of a coil to the opposite side of the coil is called the *span* of the coil and is usually given in terms of the number

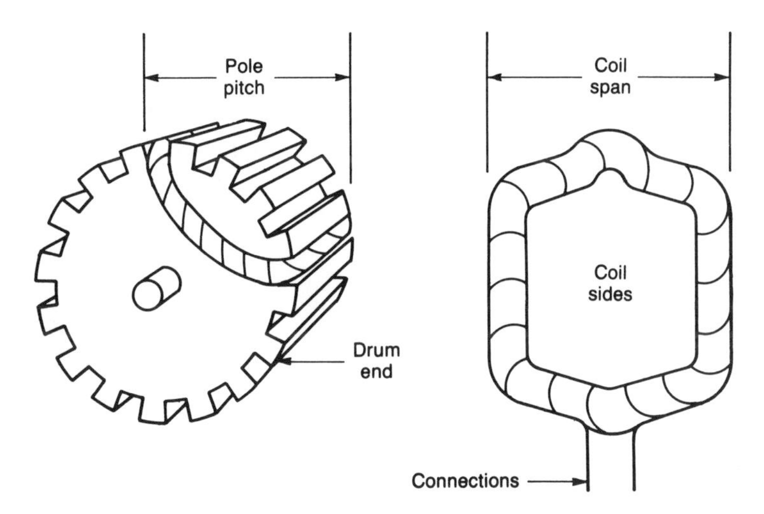

Figure 2–3 Formed armature coil. (Courtesy U.S. Navy.)

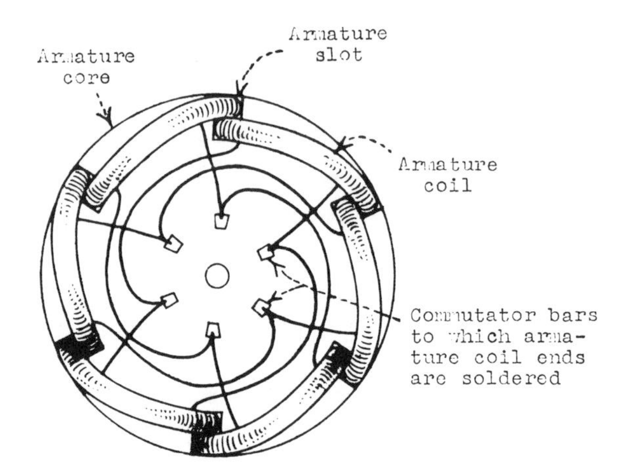

Figure 2–4 Direct-current motor armature with two-layer winding. (Courtesy U.S. Navy.)

of armature slots between them. It is generally equal to the peripheral distance between the centers of adjacent poles; this distance is called the *pole pitch*. In some instances, to save conductor expense, the span of the coil may be less than the pole pitch and is called *fractional pitch*. In direct-current motors, the armature windings usually consist of two-layer windings, one side of each coil occupying the top of one slot and the bottom of another (Figure 2–4).

Windings, having groups of coils under similar pairs of poles (at any instant having equal voltages) connected in parallel by the brushes are referred to as *lap windings* (Figure 2–5a). When these groups of coils are connected in series by the brushes, they are referred to as *wave windings* (Figure 2–5b). The winding names derive from their appearance in the winding diagrams: the coils in parallel appear to ''lap'' each other (Figure 2–5a); those in series appear as ''waves'' undulating from one side to the other (Figure 2–5b). The essential difference is that the lap winding has high-current, low-voltage characteristics, whereas the wave winding has low-current, high-voltage characteristics.

TYPES OF MOTORS

The field and armature coils of direct-current motors may be connected in three basic ways, denoted by their nomenclature: the shunt motor, the series motor, and the compound motor.

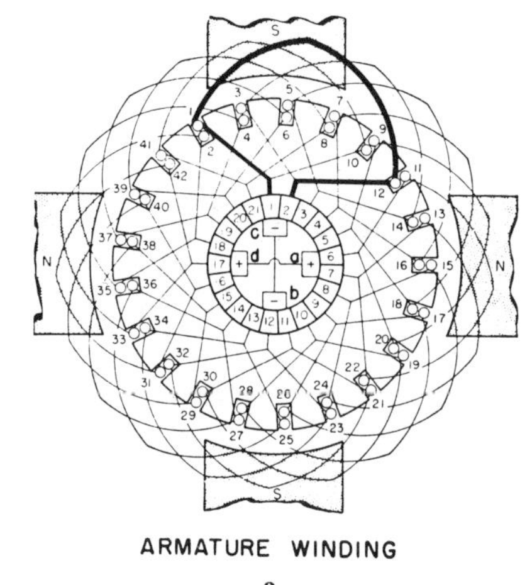

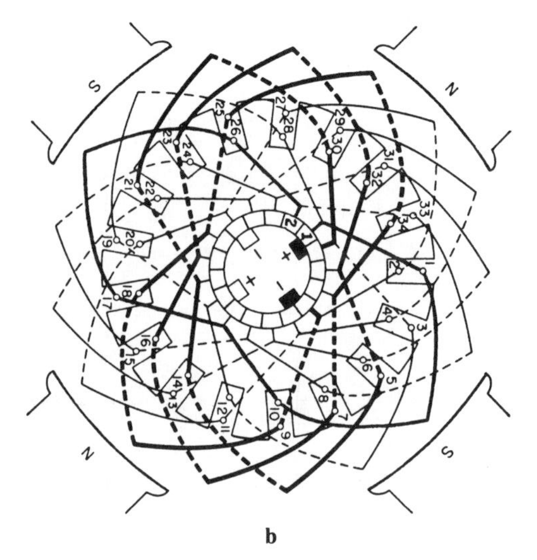

Figure 2–5 Direct-current motor armature windings: (a) lap winding: groups of coils connected in parallel by the brushes; (b) wave winding: groups of coils connected in series by the brushes. (Courtesy U.S. Navy.)

Shunt Motor

In this type of motor, the field coils (electromagnets of many turns of small-size wire) are connected in shunt (or parallel) with the armature (rotor), both connected across the incoming supply line (Figure 2–6a). If the supply voltage is constant, the current through the field coils, and consequently the field flux, will be constant. Hence the torque varies directly as the current through the armature, and the speed of the motor remains essentially constant as the load imposed on the motor varies, increasing or decreasing with its rated

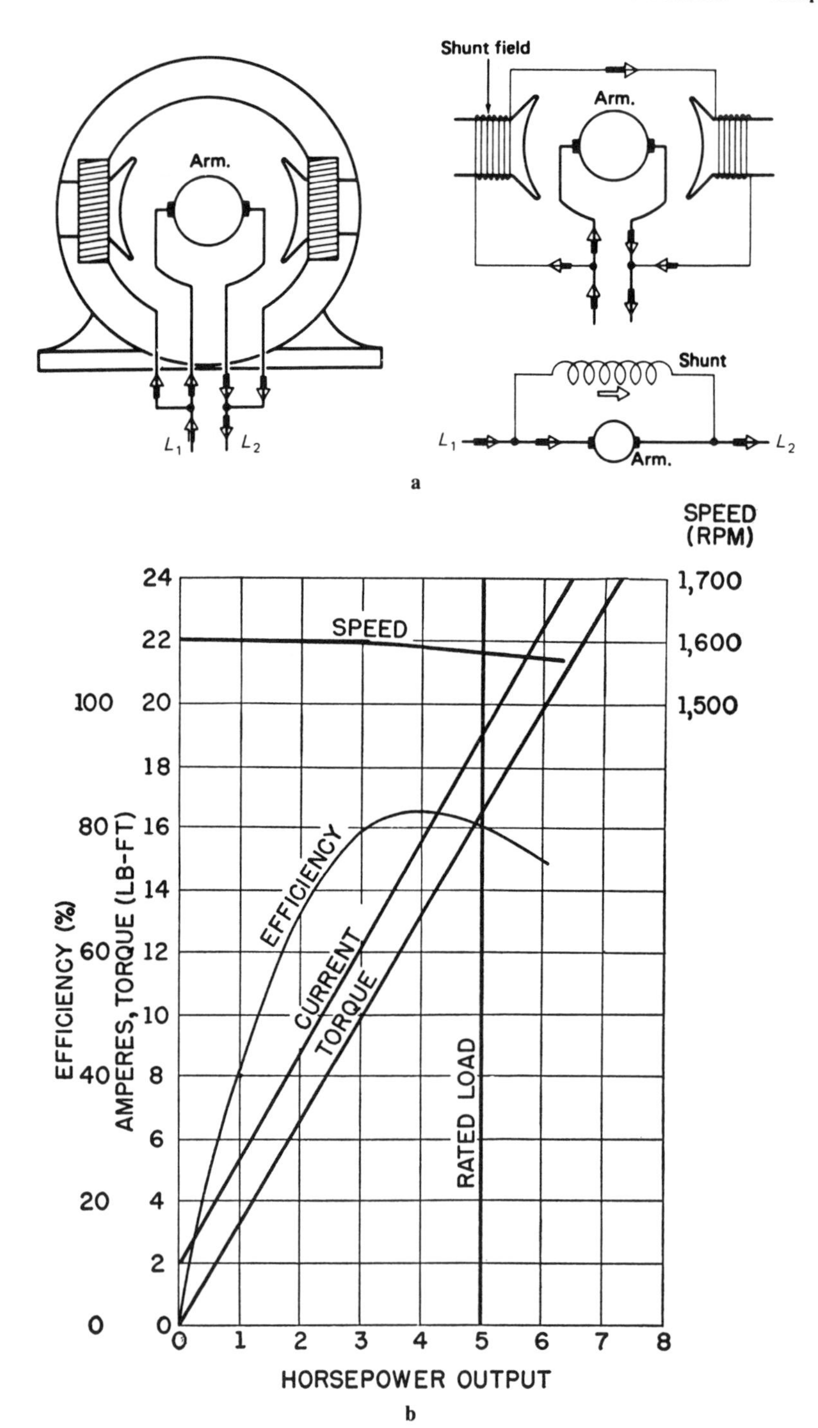

Figure 2–6 (a) Schematic diagram: (b) typical characteristic curves of a shunt motor. (Courtesy U.S. Navy.)

load (Figure 2–6b). Shunt-wound motors are used to operate machines and tools in shops and factories where constant speed is desirable. Efficiency increases rapidly with load as the losses in the field magnets are essentially constant and become a smaller percentage of total losses as the losses in the armature are relatively high.

Series Motor

In this type of motor, the field coils (electromagnets of relatively few turns of large-size wire) are connected in series with the armature (Figure 2–7a). The series field strength or flux is proportional to the armature current. The torque developed by the motor is therefore proportional to the current in the field coils and in the armature; that is, it is proportional to the square of the current. If the current is doubled, it is doubled in both the field and armature, and the torque is therefore quadrupled. The speed decreases rapidly as the load is increased (Figure 2–7b).

As indicated in Figure 2–7b, the torque increases with load. When the load on a series motor is increased, the speed and the counter EMF decrease and the armature current and field strength increase. There is, therefore, an increase in torque with decrease in speed, with the result that the increased load on the motor is limited by the decrease in speed. The series motor is accordingly used where a large starting torque is required, as in cranes, hoists, motor vehicle starters, traction equipment, blowers, and similar applications.

If the load on a series motor is very small or removed or should be reduced markedly, the speed may become very high and continue to accelerate to the point where the motor may destroy itself. An increase in speed allows the counter EMF to increase at the same time the field strength decreases; the effect of the weak field is greater than the effect of the counter EMF, so the motor tends to speed further until some load acts to stabilize the speed at a higher value or the machine breaks down. A series motor should therefore always be geared or direct connected to the load, not through a belt or chain which may break or slip its pulley or sprocket, suddenly removing the load from the motor.

Compound Motor

This type of motor is a combination of the shunt and series types of motors; one coil on the field electromagnet is in series and another in parallel with the external circuit. In most cases, the series winding is connected as shown in Figure 2–8a, so that its field is added to that of the shunt winding; such motors are called *cumulative compound motors*. If the series winding is connected as shown in Figure 2–8b, so that its field opposes that of the shunt winding, the motor is referred to as a *differential compound motor*. Generally, the ampere-turns of the shunt coil will always be greater than the ampere-turns of the series

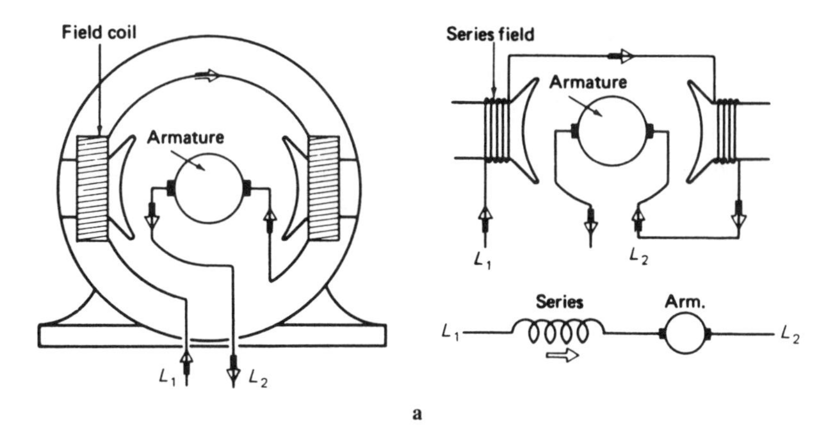

a

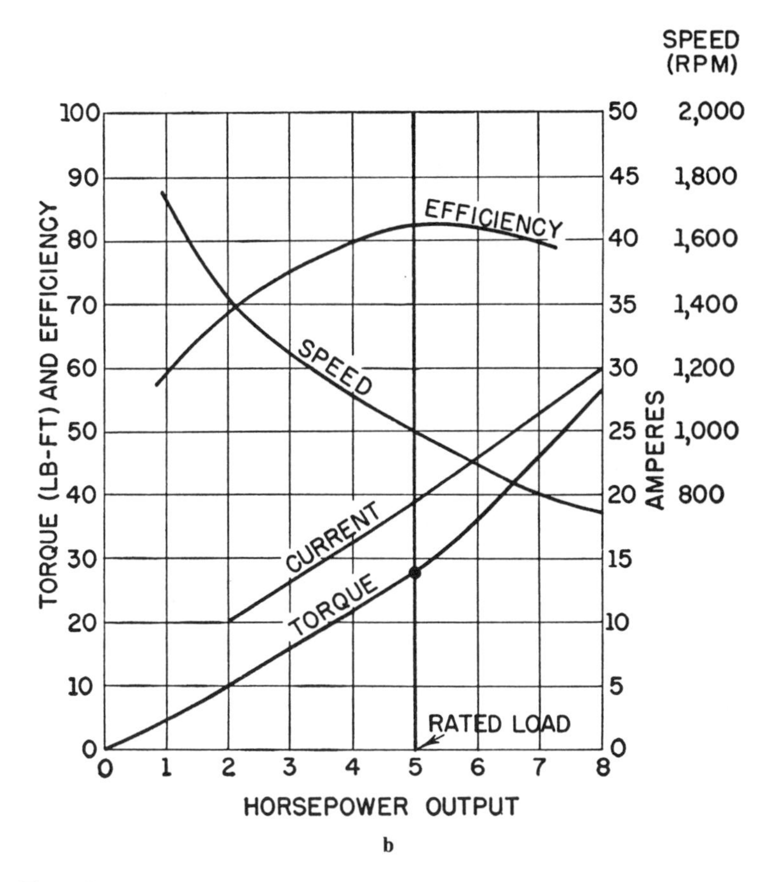

b

Figure 2–7 (a) Schematic diagram: (b) typical characteristic curves of a series motor. (Courtesy U.S. Navy.)

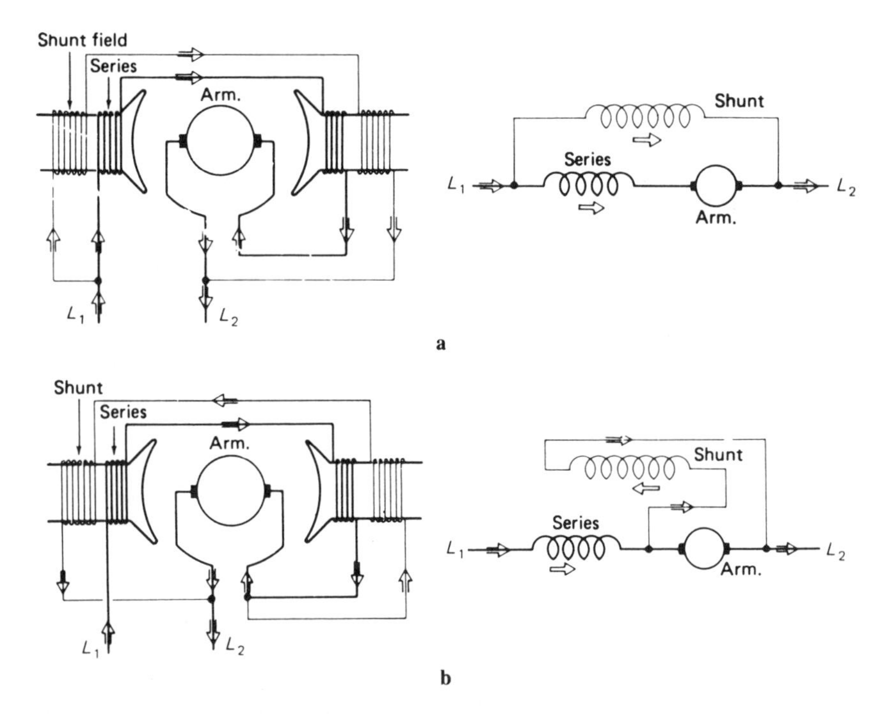

Figure 2–8 Types of compound motors: (a) cumulative; (b) differential. (Courtesy U.S. Navy.)

coil. Their characteristics compared to those of a shunt motor are shown in Figure 2–9.

Cumulative Compound Motor. In the cumulative compound motor, as the load is added the speed decreases more rapidly than it does in the shunt

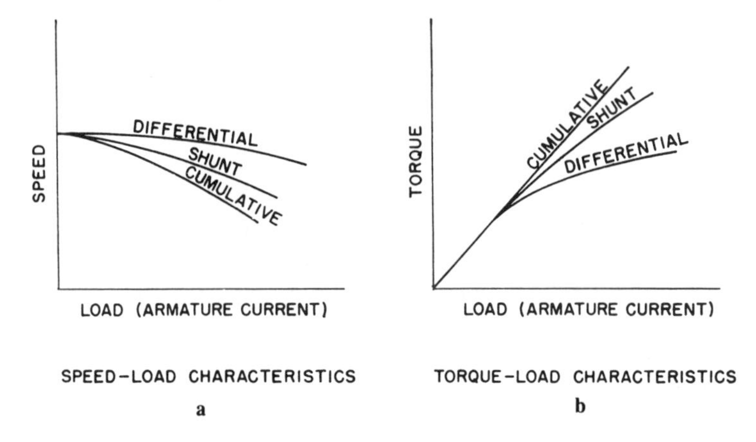

Figure 2–9 (a) Speed–load and (b) torque–load characteristics of shunt and compound motors.

motor but not as rapidly as in the series motor. The increased current necessary to carry the added load also increases the strength of the series field. To increase the load current flow, the counter EMF in the armature must decrease; but in a series motor, the decrease in speed allows the counter EMF to decrease at the same time the field increases. This adds to the flux produced by the shunt field, which is essentially constant.

As the torque developed varies directly as the product of the armature current and the field flux, the cumulative compound motor has a greater starting torque than does the shunt motor under the same conditions, but a lower starting torque than a series motor.

The characteristics of this type of motor depends on the ratio of the ampere-turns of the shunt winding to the ampere-turns of the series winding: the greater the ratio, the more nearly it approaches the characteristics of a shunt motor; the lesser the ratio, the characteristics approach that of a series motor.

When there is a decrease in the load on this type of motor, the speed tends to increase, as does the counter EMF. The current in the armature decreases as does the flux produced by the series coil; the greater part of the field flux is produced by the shunt field coils. The compound motor will therefore have characteristics similar to the shunt motor under those conditions and, unlike the series motor, will have a definite no-load speed.

When load is increased on this type of motor, there is an increase in the total flux and a greater proportional increase in torque than in the current flowing through the armature. For a given increase in torque, this type of motor requires less armature current than does the shunt motor, but greater than the series motor.

In some cases, where a good starting torque is required, the cumulative series type of motor is desirable. When the motor comes up to speed, the series winding may be short circuited. The motor then operates as a shunt motor with its better speed regulation characteristics.

This type of motor is used where high starting torque is necessary, where relatively large changes in speed can be tolerated, and where load may be safely reduced or removed; they are, therefore, used for hoists, punches, shears, and similar applications.

Differential Compound Motor. In this type of motor, the series field is in opposition to the shunt field, and hence as the load and armature current increase (as does the current in the series field), the resultant field flux decreases. The decrease in field flux will tend to have the speed increase. The increase in load and armature current, however, will tend to increase the counter EMF, which, in turn, will tend to reduce the speed. If both these actions vary in the same proportion, the speed will tend to remain constant. Indeed, if the series field strength is increased (by adding additional turns), countering even further the shunt field strength, it is possible to have this type of motor actually increase its speed as the load is increased.

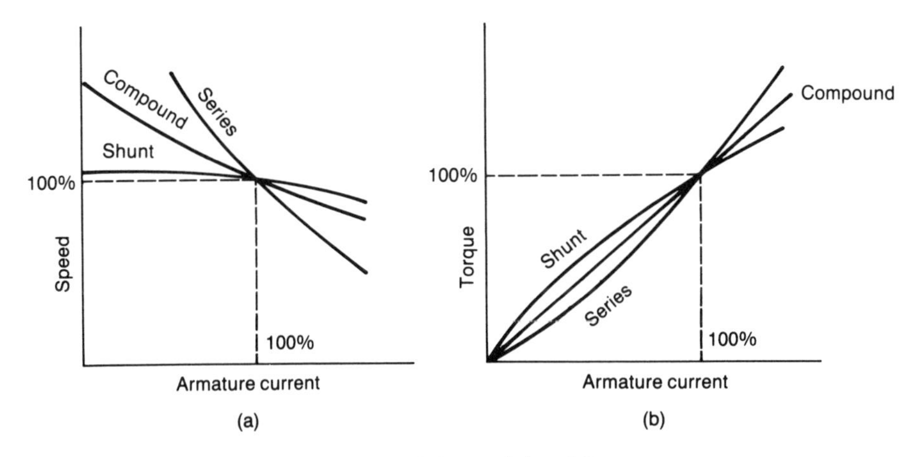

Figure 2–10 Comparison of characteristics of direct-current motors.

In this type of motor, to obtain the same increase in torque as does a shunt motor, a greater line and armature current is required than in the shunt motor. This is because this increase in current reduces the field flux. As the torque depends on the product of the armature current and field strength, the armature current must increase disproportionately as the field strength decreases in order for this product to increase. If the load current becomes excessively high, the motor will tend to run as a series motor and exhibit its characteristics as to speed and torque.

Under normal loads, this type of motor has the characteristics of a shunt motor: low starting torque and good speed regulation. Because of the undesirable features, especially during overloads, this type of motor is not widely used.

A comparison of the speed and torque characteristics of the shunt, series, and compound motors is shown in Figure 2–10.

COMMUTATING POLES

To maintain a torque in one direction when the coils of the armature (rotor) are moving under alternate poles, the current flowing in them must be reversed periodically when the brushes short circuit the coils. When the fields of the stator and rotor interact, the resultant field is tilted in a direction opposite the rotor rotation. This changes the location on the rotor where the magnetic fields in each half of the rotor come together (Figure 2–11), a neutral point where no flux is cutting the conductors. If the brushes are placed here, no sparking will occur. Since the current in the armature (rotor) changes as the load on the motor changes, the neutral point will move and the brushes must be shifted accordingly to maintain sparkless commutation. This is not always practical.

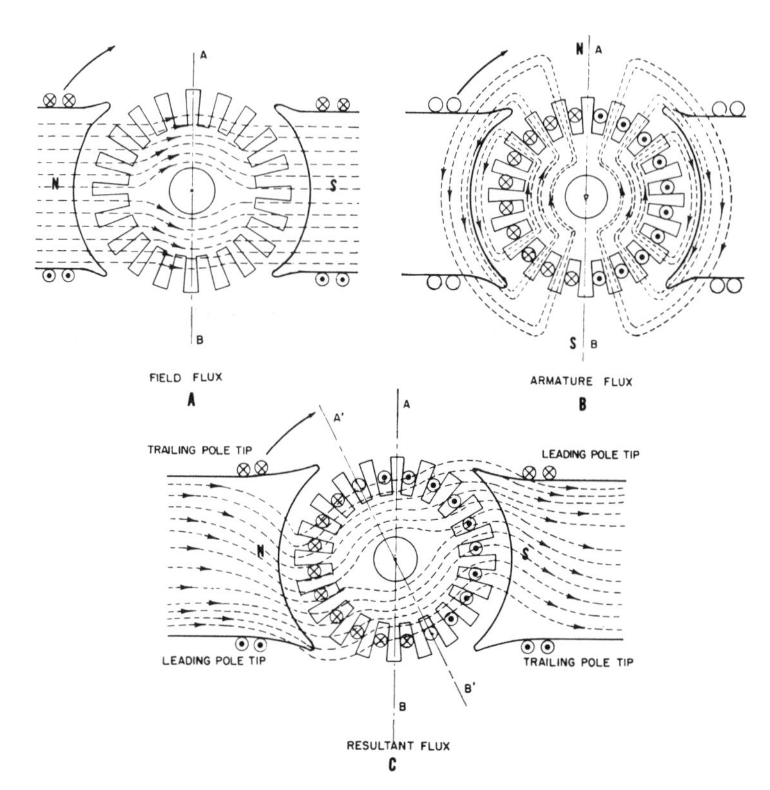

Figure 2–11 Magnetic fields in a direct-current motor. (Courtesy Westinghouse Electric Co.)

As a result of the changing current, a voltage is self-induced in the commutator short-circuited coil that tends to keep the current flowing in the same direction, as indicated by the arrow in position 1 of Figure 2–12. As the coil moves to position 2, the change in direction of the current, the self-induced

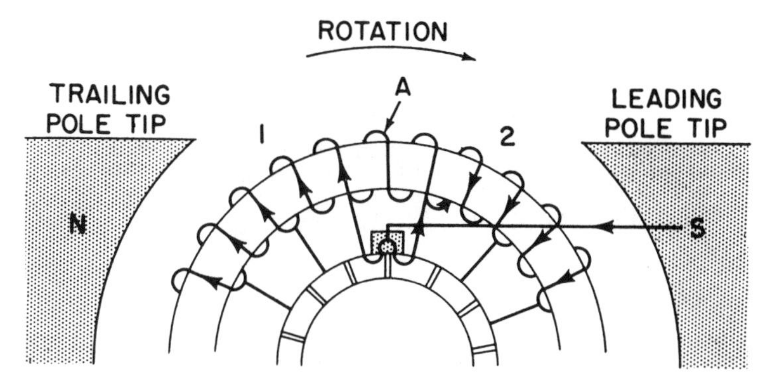

Figure 2–12 Commutation in a direct-current motor; note change in direction of armature current between points 1 and 2. (Courtesy U.S. Navy.)

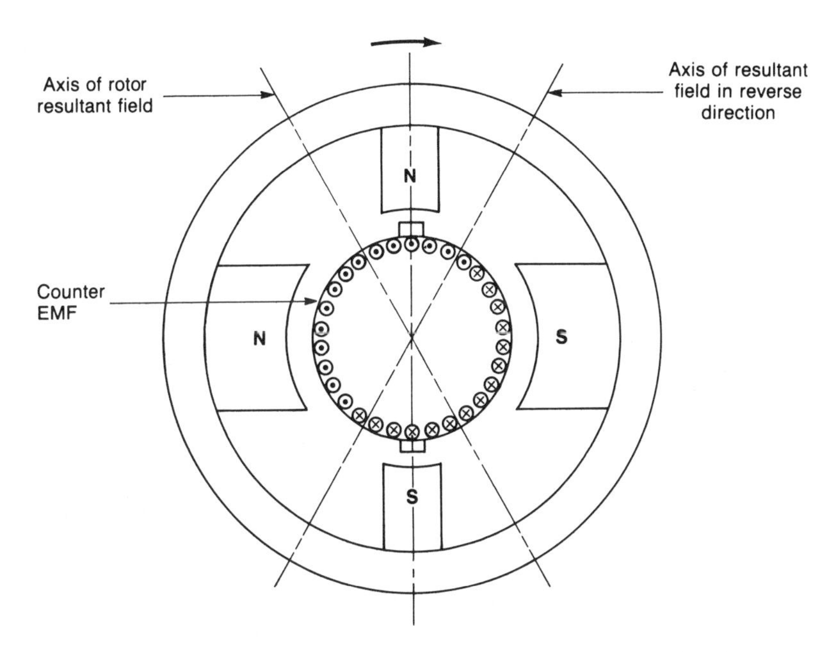

Figure 2–13 Effect of interpoles on armature counter EMF. (Courtesy U.S. Navy.)

voltage again opposes the increase in current. The induced voltage must therefore be opposed if sparking at the commutator is not to take place.

To oppose this induced voltage, another voltage must be introduced in the coil. This may be done by placing another smaller pole between the main poles of the motor; these are called *commutating poles* or *interpoles*. The polarity of this interpole has the same polarity as the main pole back of it in the direction of rotation. The winding on the interpole is in series with the armature current so that its field will vary as the armature current varies. Only the coils under the main poles contribute to the armature counter EMF that controls the armature load current and the characteristics of the motor. The field created by the interpoles will tend to create a self-induced voltage of the coils short circuited by the brushes, tending to reduce or eliminate sparking of the commutator (Figure 2–13).

STARTERS

If a motor at standstill is connected directly to its source of supply, an extremely large current would flow since the resistance of the armature is relatively very low. It is necessary, therefore, to place a resistance in series with the armature to hold the initial current to safe values until the armature begins to turn and

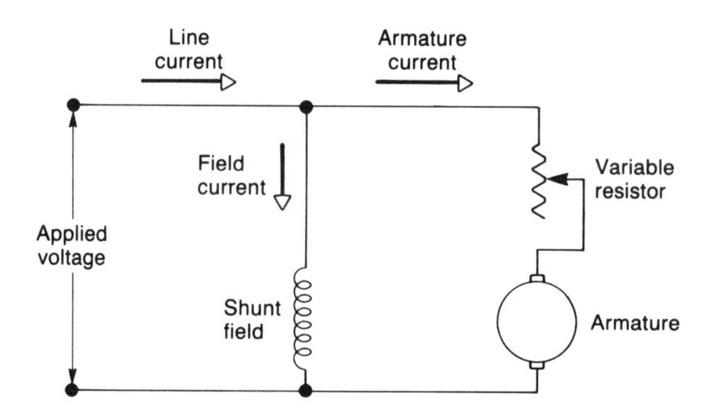

Figure 2–14 Starter and speed control by means of armature series variable resistor.

generate a counter EMF to oppose the applied voltage; in turn, this will reduce the armature current (Figure 2–14). As the armature speed increases to normal, the external resistance can be reduced, allowing full voltage to be applied across the armature.

The armature speed and counter EMF rise gradually while the external resistance is reduced proportionally as the motor approaches normal speed while carrying its load, and full voltage is applied across the armature.

The external resistances capable of being reduced gradually constitute a starter, connected in series with the armature. These starters may be operated (cut out gradually) manually or automatically. In the manually operated starter, how well the magnitude of the starting current is controlled depends on the skill and experience of the operator in manipulating the starter control handle or lever.

Example:

From Ohm's law (Chapter 7),

$$\text{armature current} = \frac{\text{applied voltage—counter EMF}}{\text{resistance of armature}}$$

At standstill, the counter EMF is zero. Assume a shunt motor with an applied voltage of 100 volts and an armature resistance of 0.1 ohm the current flow in the armature will be

$$\frac{100 \text{ volts}}{0.1 \text{ ohm}} = 1000 \text{ amperes}$$

This value will cause so much heat that insulation may fail and there is some possibility that the motor will be destroyed. It is desired to hold the starting current to a maximum of 100 A, so a resistance must be added in the armature circuit; then

$$\text{armature current} = \frac{\text{applied voltage—voltage drop in armature}}{\text{armature resistance} + \text{starting resistance}}$$

$$100\text{ A} = \frac{100\text{ volts} - (100\text{ amperes} \times 0.1\text{ ohm})}{0.1\text{ ohm} + \text{starting resistance}}$$

$$= \frac{90\text{ volts}}{0.1 + R_s}$$

Solving for R_s, we find that the starting resistance is 0.9 ohm.

As the motor begins to speed up, the counter EMF developed will be greater—say, 40 volts. Then

$$100\text{ amperes} = \frac{100\text{ volts} - [(100 \times 0.1)\text{ drop} + 40\text{-volts counter EMF}]}{0.1\text{ ohm} + \text{starting resistance}}$$

$$= \frac{100 - 50}{0.1 + R_s} = \frac{50}{0.1 + R_s}$$

Solving for R_s yields a starting resistance of 0.4 ohm.

Hence, to hold the armature current to 100 amperes, the starting resistance can be lowered to 0.4 ohm. If the starting resistance is maintained at a higher value, say 0.5 ohm, the armature current will then be

$$\text{armature current} = \frac{50}{0.1 + 0.5} = \frac{50}{0.6} = 83.33\text{ amperes}$$

On the other hand, as the motor attains its normal speed, the counter EMF will increase to, say, 90 V and the starting resistance will cut out entirely; then

$$\text{armature current} = \frac{100 - 90}{0.1 + 0} = \frac{10}{0.1} = 100\text{ A}$$

In the automatic type of starter, the external resistance is cut out gradually by a variety of devices but which are of two basic types: one actuated by a timer, which reduces the resistance at a uniform rate regardless of the load imposed on the motor; the second actuated by a voltage-sensitive relay, which reflects the counter EMF generated in the armature and the external resistance is reduced accordingly. The rate at which the resistance is reduced will vary as the load is light or heavy. Starters will operate with shunt, series, and compound motors.

SPEED CONTROL

The speed of a direct-current motor varies directly as the counter EMF in the armature and inversely as the field strength. Hence the speed may be varied by:

1. Inserting a variable resistance in series with the armature circuit
2. Inserting a variable resistance in series with the field circuit

Armature Counter EMF

The circuit containing a variable resistance in series with the armature is the same as that of the starter circuit described earlier, except that the resistance inserted must be able to carry a larger current indefinitely while the starter resistance is inserted in the circuit only a short time. The losses in the resistance are appreciable, especially at low speeds, with a consequent reduction in efficiency. This makes this type of control undesirable, especially if the motor is to operate for long periods of time (Figure 2–14).

Field Strength

The speed of a direct-current motor can also be varied by varying the strength of the field. This can be done by inserting a variable resistance in the field circuit (Figure 2–15). If the field strength is increased, the speed of the motor is decreased; if the field strength is decreased, the speed of the motor is increased.

For example, if the inserted resistance value is increased, the field current is reduced, as is the field flux. This causes a reduction in the counter EMF and an increase in current in the armature; in turn, this causes an increase in the force on the armature and an increase in its speed. If the resistance value is decreased, opposite actions take place and the speed of the armature decreases.

As the current in the field circuit is relatively small compared to the armature current (except for the series motor), losses in the inserted resistance are comparatively small and have less effect on efficiency. Hence this type of control is preferable for most motor applications.

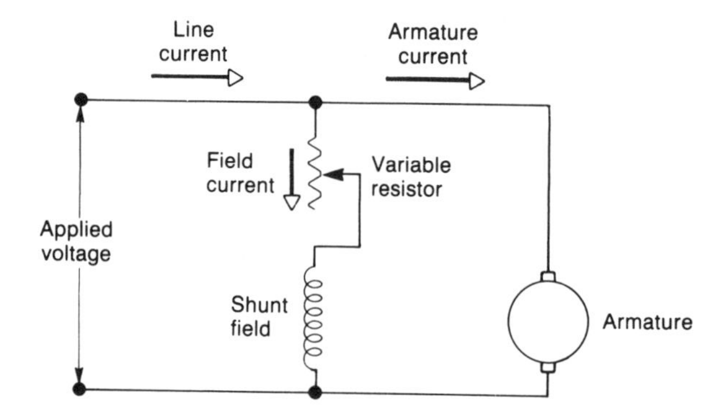

Figure 2–15 Speed control by means of field series variable resistor.

LOSSES AND EFFICIENCY

Losses

Losses in a motor result from the rotation of its movable parts and from the flow of electricity through its conductors; these are generally classified as *mechanical losses* and *electrical losses*, and all are manifested as heat.

Power is expended in overcoming friction at the bearings when the rotor turns; there is also an initial power surge necessary to overcome the inertia when the rotor begins to rotate from a standstill. There is also a loss caused by the brushes making contact with the rotating commutator. The rotor also has to overcome the resistance of the air (in the air gap), resulting in *windage loss*.

Electric power is lost in the conductors of both the field and armature conductors, generally known as *copper losses*. The magnetic fields set up by both the field electromagnets and the armature cut the metallic parts of the motor, including the field pole pieces, the armature core, and the frame of the motor. Eddy currents are set up in these, together with the hysteresis effect of the varying magnetic fields. The losses caused by the action of the magnetic fields are referred to as *iron losses* and are given off as heat.

Such losses vary as the square of the current flowing and the resistance of the current path, so-called I^2R *losses* (see Chapter 8). Since it is difficult to measure the eddy currents developed or the resistance of their paths, these losses (which include the friction and windage losses mentioned earlier) are usually determined by subtracting the copper losses from the total losses of the motor. Their losses are usually rather small compared to the copper losses and are generally considered as being essentially constant in value (Figure 2–16).

Electric power is lost (also in the form of heat) in the conductors of both the field and armature, generally known as copper losses. The field windings usually consist of relatively small size wire of many turns and carry comparatively small currents. Armature conductors generally consist of large conductors of relatively few turns and carry large currents.

Example:

The field windings of a motor consist of 1000 feet of No. 12 copper wire having a resistance at 20°C (77°F) of 1.62 ohms and carries a current of 10 amperes. The armature winding consists of 200 feet of No. 6 copper wire having a resistance of 0.403 ohms per 1000 feet at 20°C, or a value of 0.081 ohms and carries a maximum current of 100 amperes. The heat generated by the losses will affect the resistance of the conductors; a 2.5°C increase in temperature will increase the resistance approximately 1%. Assume a 50°C temperature rise; then

$$\begin{aligned}\text{field winding loss} &= 10\ \text{amperes}^2 \times (1.62 \times 120\%)\ \text{ohms}\\ &= 100 \times 1.944 = 194.4\ \text{watts}\end{aligned}$$

$$\begin{aligned}\text{armature winding loss} &= 100\ \text{amperes}^2 \times (0.08 \times 120\%)\ \text{ohms}\\ &= 10{,}000 \times 0.0972 = 972\ \text{watts}\end{aligned}$$

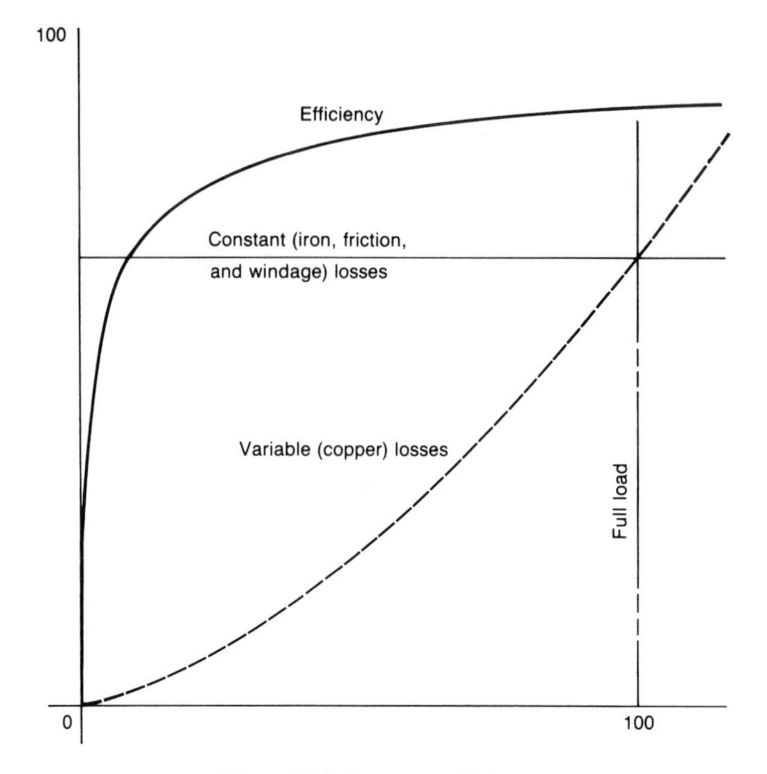

Figure 2–16 Losses and efficiency.

Total copper losses will then be 194.4 + 972 = 1166.4 watts. If the iron loss is taken at approximately 100 watts, the total losses = 1166.4 + 100 = 1266.4 watts or, say, 1250 watts.

The total losses include both the mechanical and electrical losses and will vary with the load applied to the motor, the copper loss generally being the largest and most variable of all of the components.

Efficiency

Because all machines have some losses, their efficiencies will never be 100%; that is, the output power will never be the same as the input power. Usually expressed as a percentage, *efficiency* is the ratio between output and input power:

$$\text{efficiency} = \frac{\text{output}}{\text{input}} \times 100$$
$$= \frac{\text{input} - \text{losses}}{\text{input}} \times 100$$

Example:

In the motor above, if the input is 110 amperes at 120 volts or 13,200 watts, the efficiency of the motor will be

$$\text{efficiency} = \frac{13{,}200 - 1250}{13{,}200} \times 100$$
$$= 90.5\%$$

Because of the prior appearance of steam engines in industry, motor outputs are rated in horsepower while inputs are rated in watts, or voltage and current in amperes. These can be changed to common units from the relationship: 1 horsepower is equivalent to 746 watts, or 1000 watts (1 kilowatt) is equivalent to approximately $1\frac{1}{3}$ horsepower. Efficiency curves are shown in Figures 2–6b and 2–7b. A general relationship between so-called constant losses (iron and friction losses) and variable losses (copper losses) is shown in Figure 2–16, together with a general efficiency curve.

NUMBER OF POLES

All of the previous discussion has been on motors of all types having two poles, one north and the other south, placed opposite, or 180°, from each other. With this arrangement, the armature or rotor will experience a force tending to make it rotate twice during each turn: once when its conductors are passing beneath the north pole, once when its magnetic field is being opposed by the magnetic field set up by the conductors in the armature, and again when the armature conductors pass beneath the south pole. Obviously, the rotation of the rotor will be subject to two spurts in each turn and the rotation will not be smooth unless the motor is equipped with an adequate flywheel, which will tend to maintain the rotation constant between the two spurts.

The rotation or motion of the rotor can be made more smooth and power output more even by adding additional poles, always in pairs of a north and a south. Hence a four-pole motor will have its rotor experience four spurts during a turn, a six-pole motor will experience six spurts, and so on (Figure 2–17).

The strength of the magnetic field associated with each pair of poles, however, may be decreased proportionally; that is, in a four-pole motor, the strength of the magnetic fields of each pair of poles will be (approximately) half of those of a two-pole motor, for a six-pole motor only one-third, and so

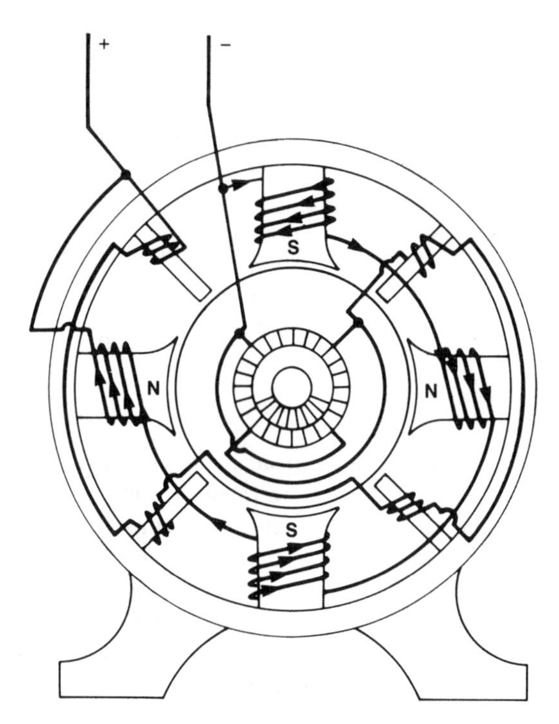

Figure 2–17 Schematic diagram of a four-pole shunt motor with interpoles. (Courtesy U.S. Navy.)

on. The ampere-turns for each pair of poles will be correspondingly less, that is, one-half as much for the four-pole motor as for the two-pole, one-third for the six-pole motor, and so on. This is reflected in fewer turns of the coils or in thinner conductors carrying a correspondingly smaller value of current.

The operation is analogous to the action found in gasoline internal combustion engines: eight smaller cylinders will produce a smoother power (driving) output than six or four larger cylinders; the horsepower output of each engine (four, six, or eight cylinders) being (approximately) the same.

Also in the discussion, the assumption has been made that the field pole pieces are stationary, attached to the frame of the motor (constituting the stator). In many applications, this is the preferred arrangement. For very large motors, however, where the armature currents may be very large, creating unusual problems with the commutator, brushes, leads, and terminals, the two elements may change places: the stator becoming the armature and the rotor, the field. Since the currents supplied to the fields are usually very much smaller, the problems associated with supplying electric current to a rotating element are much less severe. Note that the currents to the rotating field coils must be "commutated," reversing their direction, in order to maintain the same magnetic relationship between field and armature, as in the preferred arrangement for smaller-rated motors. (This arrangement is often used for large generators.)

The number of poles of a direct-current motor may affect the counter EMF induced in the armature circuit and hence the speed of the motor. The counter EMF induced in the motor armature depends on the number of lines of force per pole, the number of conductors on the face of the armature, the number of parallel paths through the armature circuit, the speed of the motor, and the number of pairs of poles. By proper design, all of these factors can be so managed that the speed of the motor can be the same, regardless of the number of pairs of poles. The final selection of the number of poles will usually be determined by space and weight limitations as well as economics.

REVIEW

- The torque of a motor depends on the strength of both the stationary magnetic field and the magnetic field produced by the current flowing in the armature (rotor).
- To keep the two fields always in a position to repel each other (to maintain rotation), it is necessary to reverse the direction of the current in the armature (rotor) twice in each revolution of the rotor; this is accomplished by the commutator (Figure 2–1).
- There are three methods of connecting the windings of the field and armature in a direct-current motor: the windings of the two fields connected in parallel across the incoming supply line (shunt motor); the two fields connected in series (series motor); and the stationary field divided in two parts, one in parallel with the armature, the other in series with the windings of the armature (compound motor) (Figures 2–6 to 2–8).
- The shunt motor is essentially a constant-speed machine. Although the speed may be varied by varying the current in the field winding; for any given field current, the speed remains essentially the same with changes in load.
- The series motor experiences a decrease in speed with an increase in load; it has a high starting torque and can accommodate starting with heavy loads. Under light load or no load, the motor has a tendency to "run away."
- The compound motor also has a high starting torque but can tolerate large variations in speed; load may be removed in safety.
- Sparkless commutation may be obtained by the shifting (with load) of the brushes to keep them in the neutral axis of the rotor field, or by the installation of interpoles whose windings are in series with the armature

(rotor) current, placed between the main poles and generally located at the brush positions (Figure 2–13).

- Motors are generally started at lower than line voltage before full voltage is imposed on them; it may be accomplished by the insertion of variable resistance in series with the armature windings, and the same variable resistance may be used to control the speed of the motor (Figures 2–14 and 2–15).
- Losses in a motor include those caused by current in the windings of both stator and rotor, generally called copper losses; and those caused by currents induced in the rotor and stator cores by the moving magnetic fields of the armature and field cutting the poles, cores, and other metallic parts of the motor, generally called iron losses. These, and those caused by friction of the bearings, brushes, and the air (windage), are usually considered as constant losses.
- While the efficiencies of each type of motor vary somewhat with the loads imposed on them, all of them are relatively high, in the nature of 80 to 90% at rated loads.
- The number of poles of a motor is determined generally by the desired rate of speed of the motor, the smoothness of rotation of the rotor, and on the overall economics in achieving these objectives.

STUDY QUESTIONS

1. How is the field winding of a shunt motor connected to the motor armature? Of a series motor? Of a compound motor?
2. What is meant by torque? How is it expressed?
3. How does the torque of a compound motor differ from that of a shunt motor? Of a series motor?
4. What is the function of a commutator in a direct-current motor?
5. How many foot-pounds per hour make up one horsepower of work?
6. In what direction, relative to the rotation of the motor, are the brushes shifted in a noninterpole motor when the load increases?
7. What is an interpole, and what is its function?
8. How does a load increase affect the speed of a series motor? How does a load decrease?
9. What are the speed–torque characteristics of a shunt motor? Of a series motor?
10. Why is a starting resistance needed for many motors?

chapter 3

Alternating-Current Polyphase Induction Motors

INTRODUCTION

Although the concept of alternating current was known since the earlier development of direct-current generators and motors in the early 1830s, it was not until 1885 that Galileo Ferraris developed the alternating-current induction motor. This invention followed by a few years (1883) the invention of the transformer by Gaullard and Gibbs. The transformer's ability to change voltages by induction processes was demonstrated in the first transmission line, all of 17 miles long, at Cerchi, Italy, in 1886; this line raised and lowered an alternating-current voltage from 100 volts to 2000 volts and back to 100 volts. From these beginnings, alternating-current systems and motors have come to overshadow the original direct-current systems.

Alternating-current motors have certain advantages over direct-current motors: in general, they are less expensive; in many instances they do not employ commutators and brushes; and they are more reliable with fewer maintenance requirements. They may operate on polyphase or single-phase supply and at several voltage ratings. They are also classified as induction type and synchronous type; the latter are discussed separately in Chapter 4. The same general electromagnetic premises that apply to direct-current motors (Chapter

2) also apply to alternating-current motors and should be kept in mind continually.

PRINCIPLE OF OPERATION

In both direct-current and alternating-current motors, the driving torque is provided by the interaction between the magnetic fields set up by the stator and rotor. In the direct-current motor, the magnetic field is usually stationary, and the field, set up by the armature with the current-carrying conductors, rotates. The current is supplied to the armature through a commutator and brushes.

In the alternating-current induction motor, the currents in the rotor are supplied by induction (hence the name). The conductors of the rotor are cut by the stator alternating magnetic field (set up by the alternating-current supply to the field), inducing alternating current in the armature conductors. The electromagnetic interactions between the two coils also constitute transformer action and, in many ways, exhibit the same characteristic; more detailed data are provided in Chapter 10.

POLYPHASE CONNECTIONS

In three-phase induction motors, the separate phase coils of the stator may be connected in delta (Δ) or wye (Y), as shown in Figure 3–1. Both connections produce magnetic fields of each phase displaced 120 electrical degrees from each other.

In the delta connection, the winding insulation is normally subjected to a higher voltage than that of the wye connection. The delta connection is usually not grounded, so that a (single) ground on one phase will not affect the operation of the motor; two or more grounds, however, constitute a short circuit between the points of the grounds and result in improper operation or failure of the motor.

The wye connection is generally grounded at the common point, and while the winding insulation is usually subjected to a lower voltage (than the delta connection), a ground on any of the phases will adversely affect the operation of the motor.

In the diagrams of Figure 3–1, it will be noted that the voltages imposed on the individual coils may vary from the maximum full-line voltage when all of the phase coils are connected in parallel in the delta connection to a minimum when the phase coils are connected in series in the wye connection. The currents flowing in the individual coils may vary from the maximum full-line current when connected in series in the delta connection to a minimum when the phase coils are connected in parallel in the wye connection. The total ampere-turns for each of the arrangements is the same for the delta connection and for the wye connection. If the individual coils in both the delta and wye connection are assumed identical, the total ampere-turns for the delta connection, however, may be higher than that for the wye connection, as the voltage impressed on the coil arrangements for the wye connection is only 0.866 that of the delta connection.

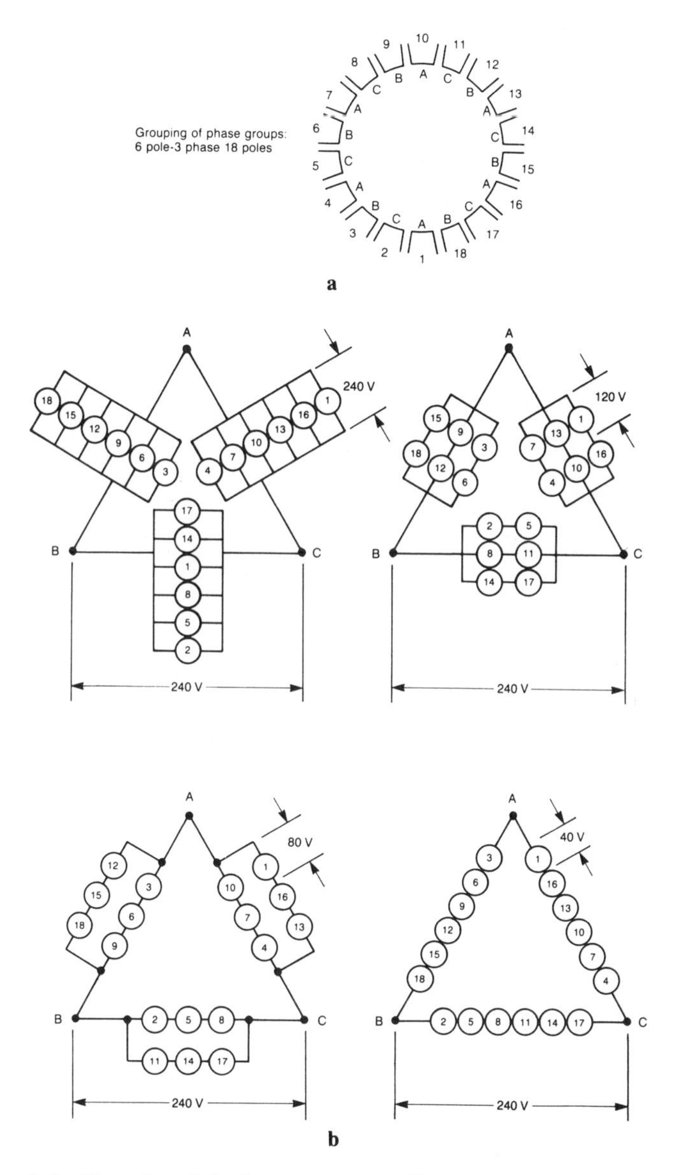

Figure 3–1 Three-phase induction motor stator coil arrangements for delta and wye connections. (continued on next page)

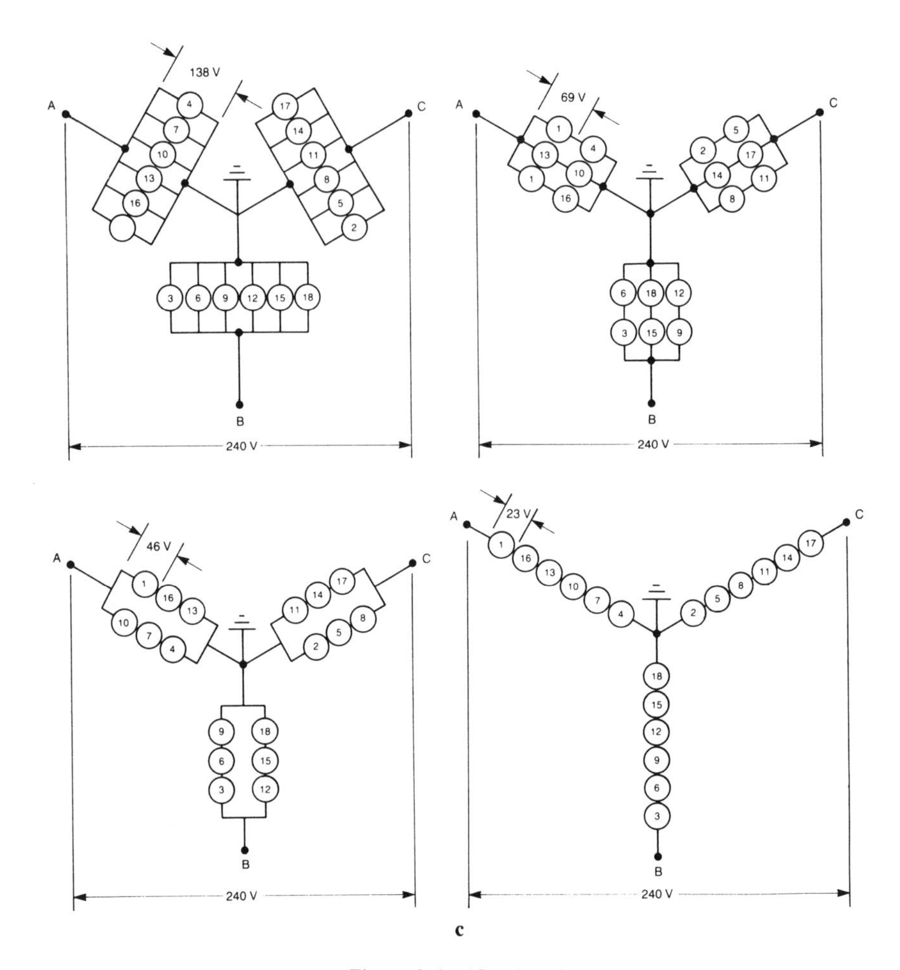

Figure 3–1 (*Continued*)

ROTATING MAGNETIC FIELDS

In the alternating-current induction motor, the magnetic fields of both the field (stator) and the armature (rotor) rotate.

Stator

Assume a three-phase alternating-current supply to the field or stator. Each phase will produce a magnetic field around its conductors that will fluctuate as the alternating-current supply varies. This variation may be portrayed graphically; the shape or form of the curve produced is known as a *sine wave* (Figure 3–2). The strength and direction of the magnetic field produced will also vary as the current variation or as a sine wave.

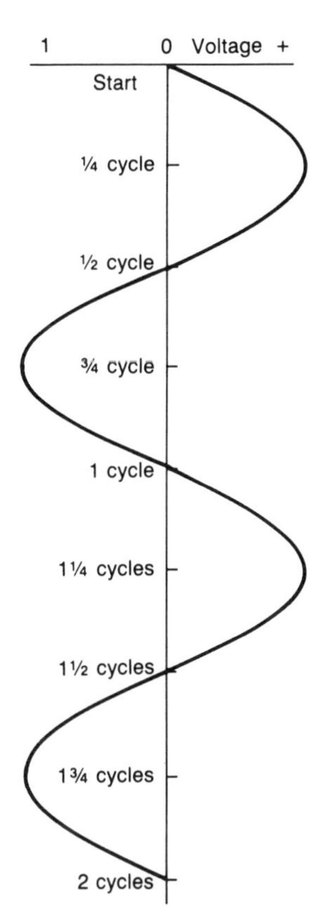

Figure 3–2 Voltage variation: alternating-current system, sine-wave shape.

In a three-phase supply, each phase produces a current that varies as a sine wave, but each phase electrically is displaced one-third of a revolution or 120° apart, shown graphically in Figure 3–3. If these three phases supply the coils for the field of the motor situated 120° apart, as shown in Figure 3–4, each coil will produce its own magnetic field whose strength and direction will vary in a sine-wave fashion.

At any instant, the space distribution of the flux caused by any one phase will be determined by the arrangement of the winding (the number of conductors adjacent to each other). If the space distribution of the flux produced by each phase is sinusoidal (follows the characteristics of a sine wave), the basic time variation of the flux in the air gap for the three phases combined will produce a rotating magnetic field revolving at a speed constant in value and depending on the frequency of the alternating-current power supply. This is referred to as the *synchronous speed.*

Synchronous Speed. This is illustrated in Figure 3–5. Figure 3–4 is a simplified diagram of a two-pole stator field winding of a wye-connected three-phase motor. Figure 3–5 shows the three-phase currents in their time-phase

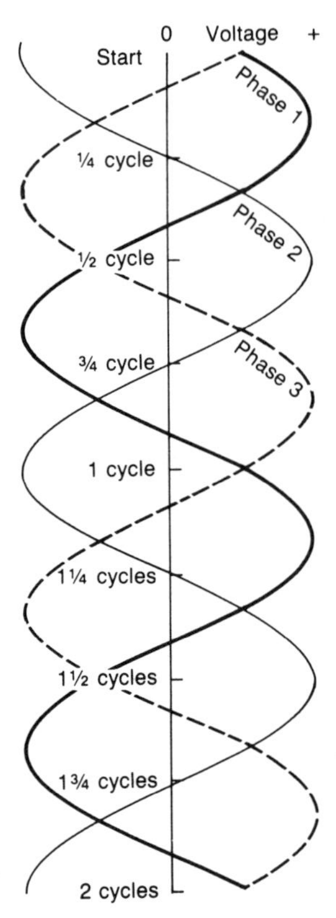

Figure 3–3 Voltage variations: three-phase alternating-current system.

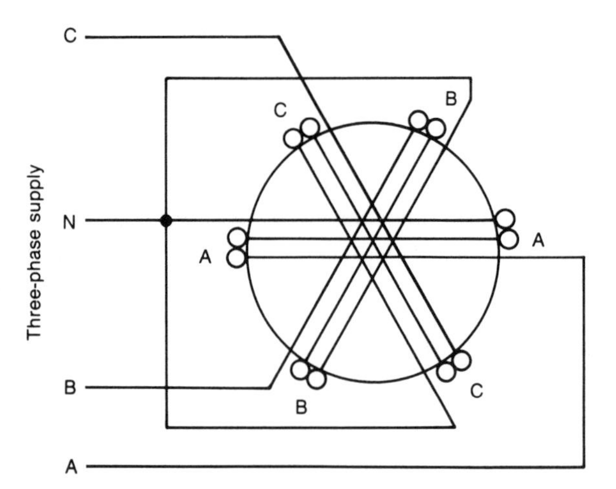

Figure 3–4 Simplified diagram of three-phase wye-connected stator.

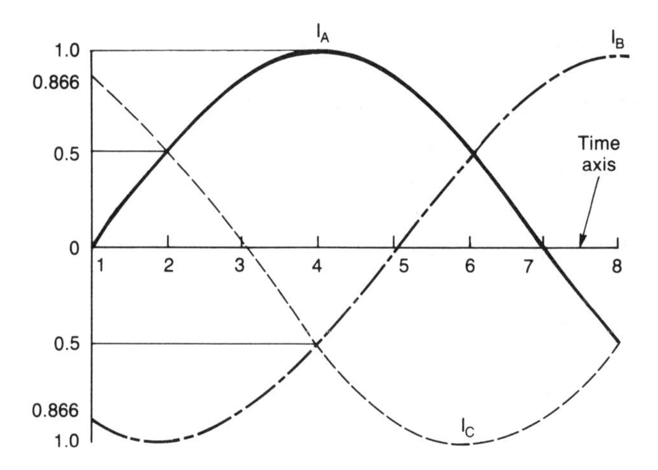

Figure 3–5 Current relationships of three phases of alternating-current system.

relation to each other. Figure 3–6 shows the direction of the currents in the three phases at several 30 degrees successive intervals as indicated, and Figure 3–7 shows the fields around the conductors at the instantaneous values of currents in the phases and shows the direction of rotation of both the rotor (armature) and stator (field) magnetic fields as well as the direction of rotation of the rotor.

At instant 1, the current in phase A is zero and 0.866 of their maximum values in each of phases B and C, but opposite in direction. The current direction in phase B is toward the observer through the upper half and away from the observer in the lower half of the B phase stator coils to the midpoint or neutral of the wye connection. The current in phase C flows away from the observer in the upper half of the C phase stator coils to the neutral point. As the instantaneous direction at point 1 of the currents in phases B and C is the same and flows as shown in Figure 3–6, the magnetic field produced is as shown.

At instant 2, the current in phase B has increased to its maximum negative value and the currents in phases A and C are at half of their maximum values. The currents in phases B and C are in the same direction as at instant 1. The current in phase A at the same instant is in a direction opposite to that in phases B and C in the upper part of the coil. The currents in each half of the coil and the magnetic fields are shown at point 2, with the resultant magnetic field rotated 30° from its position at instant 1.

Continuing this analysis for instants 3 through 12 produces the current and magnetic fields shown. It will be noted that the resultant magnetic field in instant 3 has rotated another 30° in the same direction from its position at instant 2, and another 30° at instant 4 from its position at instant 3—and so on to instant 12. The magnetic field rotates around the stator, moving one degree of space for each degree of time shown in the sine curve time axis of Figure 3–5.

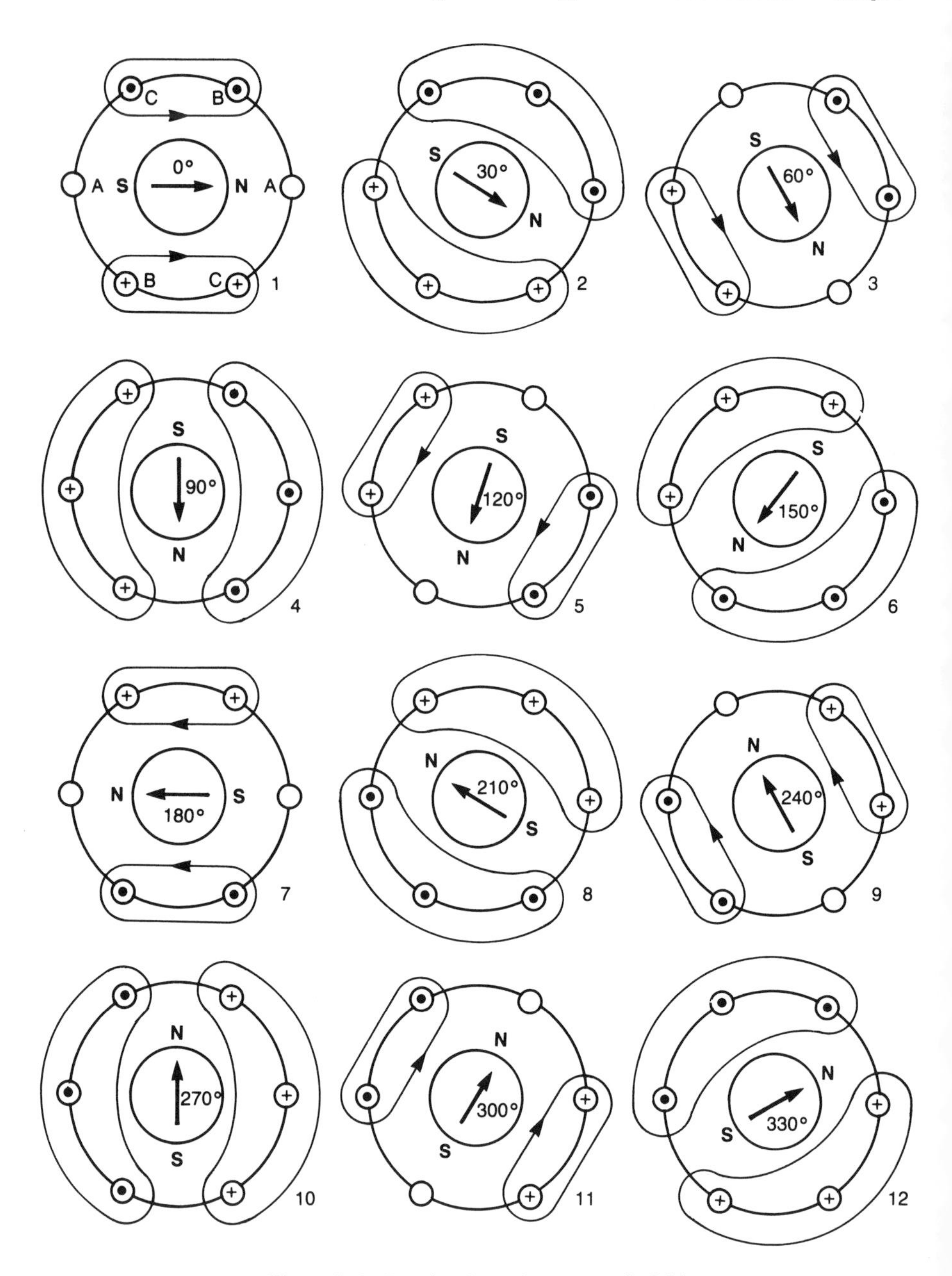

Figure 3–6 Rotating three-phase magnetic field.

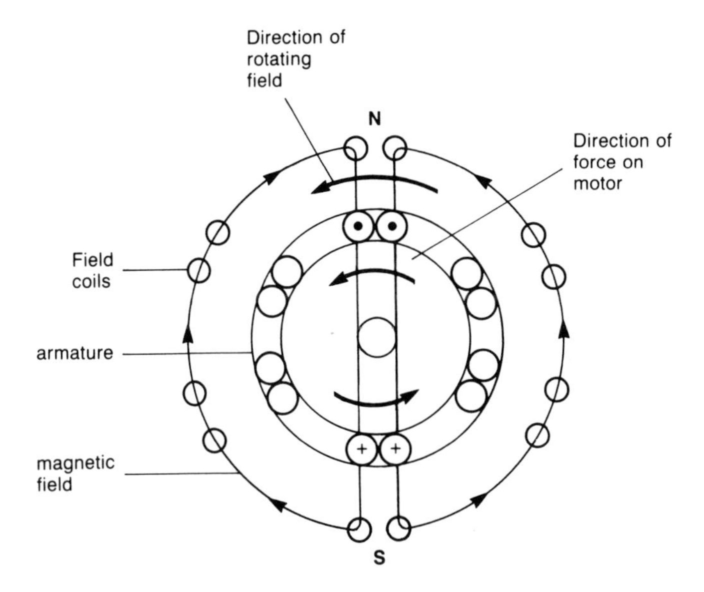

Figure 3–7 Direction of rotation of motor and field.

Reversing Field. The direction of the rotating magnetic field may be reversed by interchanging any two of the supply wires to the three terminals of the motor. In Figure 3–4, supply lines A, B, and C connect to phases A, B, and C, respectively, and each of their line currents reach their maximum values in that order. The phase sequence is A, B, C, with the resulting field rotation in one direction. If supply lines A and B are interchanged, the phase sequence becomes B, A, C and the rotating field turns in the opposite direction.

Frequency. As mentioned earlier, the speed of the rotating field varies directly as the frequency of the alternating-current supply and inversely as the number of poles. The speed of the rotating magnetic field is always independent of load changes on the motor if the frequency is held constant.

Example:

In the discussion so far, a simple two-pole three-phase stator winding was assumed. If the frequency of the supply alternating current is taken at 60 cycles per second, the speed of the rotating magnetic field; that is, its synchronous speed, is

$$\text{speed} = \frac{\text{frequency} \times 60 \text{ seconds}}{\text{number of pole pairs}}$$

$$= \frac{60 \text{ cycles/second} \times 60 \text{ seconds}}{2 \text{ poles/2}}$$

$$= 3600 \text{ rpm}$$

Increasing the number of poles lowers the synchronous speed; if the motor had four poles, the synchronous speed would be 1800 rpm. Decreasing the frequency of supply alternating current decreases the synchronous speed; if the motor was supplied at 25 cycles, the synchronous speed would be

$$\text{speed} = \frac{25 \text{ cycles per second} \times 60 \text{ seconds}}{2 \text{ poles}/2}$$
$$= 1500 \text{ rpm}$$

Rotor

The rotor of an induction motor essentially acts as the secondary of a transformer, with the stator acting as the primary. The rotating magnetic field of the stator cuts the conductors of the rotor, inducing currents in them. These currents, in turn, produce a magnetic field similar to that produced by the stator. The interaction of these two fields develops a torque that causes the rotor to turn.

It is to be noted that there is no external electrical supply to the rotor conductors; the currents in the rotor are entirely induced. The rotor conductors are connected together at each end of the rotor completing their circuits (Figure 3–8).

Motor action, illustrating the rotating magnetic field, is shown in Figure 3–7. A two-pole rotating magnetic field, revolving at synchronous speed, is (assumed) turning in a counterclockwise direction. At the instant shown, the north-pole field is cutting across the conductors on the upper part of the rotor in a right-to-left direction; the relative direction of the conductors, with respect to the field, however, is to the right. Applying the right-hand rule for generator action (Chapter 1): the thumb points to the left, the index finger points downward, and the middle finger points toward the reader, indicating that the current flow in the upper conductors is toward the observer.

For motor action, applying the left-hand rule to the conductors of the rotor, the direction of the force acting to turn the rotor is determined: the index finger points downward, the middle finger points toward the observer, and the thumb points to the right—indicating that the force on the rotor tends to turn it in a counterclockwise direction, the same as that of the rotating field.

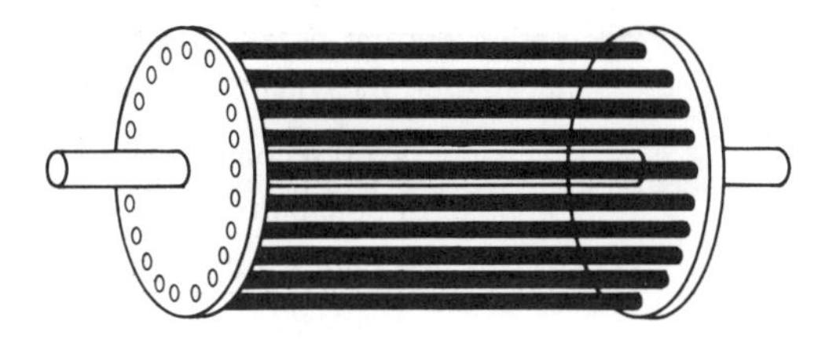

Figure 3–8 Elementary squirrel-cage rotor.

Applying the left-hand rule to the conductors on the lower part of the rotor, the force on the rotor indicated will also be in the same direction as the field, counterclockwise. Thus the forces acting on the upper and lower parts of the rotor act in harmony to turn the rotor in the same direction. And, with no load on the motor, the rotor will theoretically turn at the same speed as the speed of the rotating magnetic field, that is, at synchronous speed; this does not take into account the effect of electrical and mechanical losses, including windage losses.

MOTOR OPERATION

Slip

If the rotor could turn at synchronous speed, there would be no relative motion between the conductors of the rotor and the rotating stator magnetic field and therefore no current induced in the rotor conductors. With no rotor magnetic field to interact with the rotating stator magnetic field, no torque would be produced.

The induction motor therefore cannot rotate at exactly the synchronous speed. But because of the losses in the rotor and the load imposed on the motor, the rotor will turn at a speed below the synchronous speed. This will cause a relative motion between the conductors of the rotor and the rotating magnetic field, and currents will be induced in the rotor conductors. This current will, in turn, produce a magnetic field that will interact with the rotating magnetic field, producing a torque necessary to overcome the resistance to turning caused by the losses in the rotor and the load imposed on the motor. The difference between the speed of the rotating stator magnetic field (the synchronous speed) and the resultant speed of the rotor is known as the *slip* of the induction motor.

Rotor Frequency

The frequency of the current induced in the rotor conductors depends on the speed of the rotor with respect to the speed of the rotating stator field. One cycle of alternating current is induced in the rotor when the rotating magnetic field makes one complete turn around the rotating rotor. The frequency of the induced rotor current is directly proportional to the slip; slip is generally expressed as a percent of the synchronous speed and the frequency in cycles per second.

Example:

If the slip is assumed to be 10% and the supply frequency at 60 cycles per second, the frequency of the current induced in the rotor will be 10% of 60 cycles or 6 cycles per second. The speed of the rotor will decrease; that is, the slip of the motor will increase, as the frequency and current induced in the rotor increase in magnitude.

Torque

Torque is the turning tendency and is equal to the force produced by the rotor times the radius of the rotor; it is expressed in pound-feet.

The torque developed in any motor is caused by the interaction of the magnetic fields of its field (stator) and its armature (rotor). The magnetic fields in turn are produced by the currents flowing in the coils or conductors producing them. The resulting torque, is therefore, directly proportional to the current in the rotor and the strength of the stator field. In the induction motor, the stator field is a rotating field, and the resultant torque is in the direction of the rotating field.

Induction

Up to this point, the discussion has mentioned current induced in conductors when they cut or are cut by magnetic fields. In fact, however, a voltage is induced in such conductors and currents do not flow until a circuit is completed. When current flows in a conductor, it will set up a magnetic field that not only interacts with other magnetic fields, as described earlier, but also induces voltages in other nearby conducting material (including the conductors) producing adjacent magnetic fields (and in the conductors themselves) in which the (''induced'') current flows.

These phenomena are called *mutual inductance* and *self-inductance*. The effect of these phenomena is to cause the voltages induced and the current flow not to operate in step with each other, but out of step (or out of phase) to a greater or lesser degree depending on the characteristics of the circuits and the conditions under which they operate.

The net effect is that only a portion of the power input is effective, and this relationship between the voltage and current is known as *power factor* and is expressed in percent. When both current and voltage are acting completely together, the power factor is 100%; when acting at right angles to or displaced from each other one half-cycle, the power factor is zero. For a more thorough explanation of these phenomena, see Chapter 10.

When the motor starts and the rotor slip is 100%, the effect of the inductance between the stator field and rotor conductors is such that the power factor of the rotor current is approximately zero. As the motor accelerates to a speed approaching the synchronous speed and the rotor slip approaches zero, the power factor of the rotor current approaches approximate unity or 100%. Normal operation is between these two extremes, with the motor slip usually 10% or less.

Rotor at Rest

At the start, with the rotor at rest, the induction motor acts as a transformer with the stator field acting as the primary and the rotor as a short-circuited secondary (see Chapter 10). A counter EMF is generated in the stator winding, which limits the line current to a small value; this no-load value is called the *exciting current* and its function is to maintain the revolving field. Under this condition, the magnetic field set up in the rotor is countered by that of the rotating magnetic field, with little interaction between the two tending to turn the rotor; that is, the torque produced is essentially zero. The slip is 100% and the rotor current power factor current approximately zero.

Rotor in Motion

As the motor approaches synchronous speed, the slip will be almost zero and the voltage induced in the rotor will be nearly zero. The current flowing in the rotor will be very small producing a very weak rotor magnetic field to interact with the rotating magnetic field and hence the torque produced will also be very small. The slip approaches 0% and the small rotor current power factor approximately 100%.

These represent the two extreme conditions: when the rotor is not turning and when it is turning at synchronous speed. Normal operation, as mentioned earlier, falls between these limits and the motor speed is usually about 10% less than synchronous speed.

Load Addition

As load is applied to the motor, the rotor will slow down, but the rotating magnetic field continues at synchronous speed; hence the slip and rotor current will increase. The torque and power output will tend to increase more than the decrease in speed. The magnetic field of the rotor will increase and will oppose the rotating magnetic field, lowering its strength slightly; the counter EMF induced in the stator field will therefore also decrease slightly. This will permit the current in the rotating magnetic field coils to increase to maintain the rotating field and prevent its weakening because of the opposition from the rotor current. Because of the relatively low internal resistance (impedance) of the motor windings, a small reduction in speed and counter EMF in the stator field may be accompanied by large increases in motor current, torque, and power output.

Pull-Out Point

As load is applied, the motor will slow down and the slip increase. The interaction between the magnetic field of the rotor and the rotating magnetic field reaches a maximum when the voltage induced in the rotor and the current

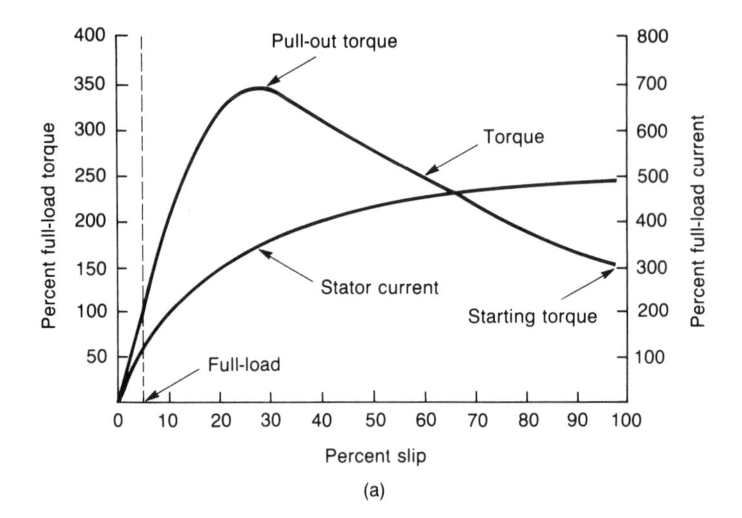

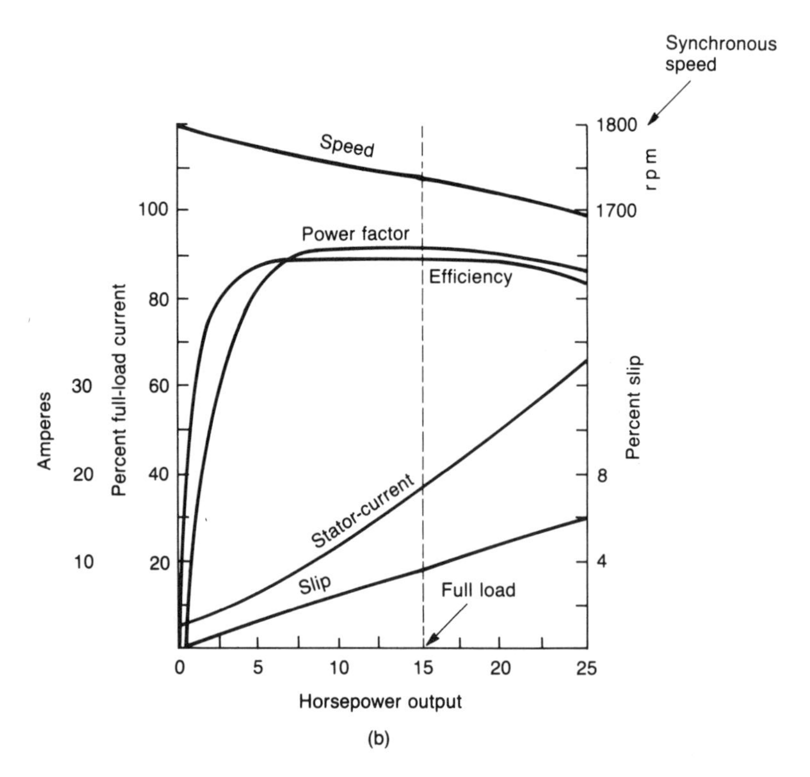

Figure 3–9 Characteristics of a squirrel-cage motor: (a) torque and current curves; (b) performance curves for three-phase squirrel-cage induction motor. (Courtesy Westinghouse Electric Co.)

flow are out of phase so that their power factor relationship reaches 70.7%. This point is called the *pull-out point*; beyond this point, the motor speed falls off rapidly with added load and the motor will stall.

Figure 3–9 portrays characteristic curves of a polyphase induction motor. When the motor is running, the rotor current is determined principally by the resistance of the rotor; the torque increases to the pull-out point as the slip increases. Beyond this point the torque decreases and the motor stalls. As the change in speed from the no load to full load is relatively small, the motor torque and the horsepower output are considered to be directly proportional (Figure 3–9).

Effect of Supply Voltage

Fluctuations in the supply voltage to the motor (actually to the stator field) affect the torque of the motor. The applied voltage determines the strength of the rotating magnetic field, which, in turn, determines the induced rotor current. As the rotor torque varies as the product of both these factors, the torque of an induction motor will vary as the square of the applied supply voltage.

The rotating magnetic field also induces a counter EMF in the stator winding (similar to the counter EMF generated in the armature of a direct-current shunt motor). The counter EMF is in opposition to the applied voltage and tends to limit the stator current. The counter EMF, however, is limited by the saturation* of the magnetic core of the stator, so that if the applied voltage is increased beyond this point, the current in the stator coils may become excessively high, causing damage to the stator of the motor.

ROTOR TYPES

The induction motor differs from direct-current motors and the synchronous alternating-current motor in that the current in its armature, usually the rotor, is produced by electromagnetic induction (hence its name) rather than by conduction. This means that electrical connection between the supply voltage and the rotor is not necessary—that commutator, slip rings, and brushes are not needed—simplifying the construction, maintenance, and operation of this type of motor. There are two distinct types of rotors: the squirrel-cage type and the wound type.

* (How many magnetic lines of force can be carried within the metal before the resistance to them becomes so high that additional lines of force will be canceled by the resistance encountered and no overall increase in current flow will occur.)

Squirrel-Cage Type

Such a rotor has its conductors, usually of copper but sometimes of aluminum, embedded horizontally lengthwise in the surface of the steel core and connected to conducting end rings or straps on each end, forming a structure that can take on the appearance of a squirrel cage, hence its name (Figure 3–8 and 3–10a). Since the resistance of the conductors compared to the steel core in which they are embedded is very low, the conductors are often not insulated, although in many instances some insulation is provided.

Rotors of this type are extremely rugged and have very low resistance. They have low starting torque but good speed regulation; starting current is large and the power factor at starting is low. (The discussion heretofore assumes this type of rotor.)

By varying the designs of the elements of the squirrel-cage rotor, its characteristics may be modified to accommodate the several types of loads that it may be called upon to serve. Squirrel-cage rotors may therefore be further classified (by the National Electrical Manufacturers' Association) as follows:

Class A: normal torque; normal starting current: low resistance of the cage; results in good running efficiency and power factor, high pull-out torque, low slip, moderately high starting torque and starting current. Generally suitable for full-voltage starting, reduced voltage to meet supply system starting current limitations (for units generally above 5 horsepower).

Class B: normal torque; low starting current: high resistance of the cage; results in efficiency, slip, starting torque same as class A, but power factor and pull-out torque slightly lower. Suitable for full-voltage starting.

Class C: high torque; low starting current: double squirrel cage; combines high starting torque (200% or more) with low starting current of class B;

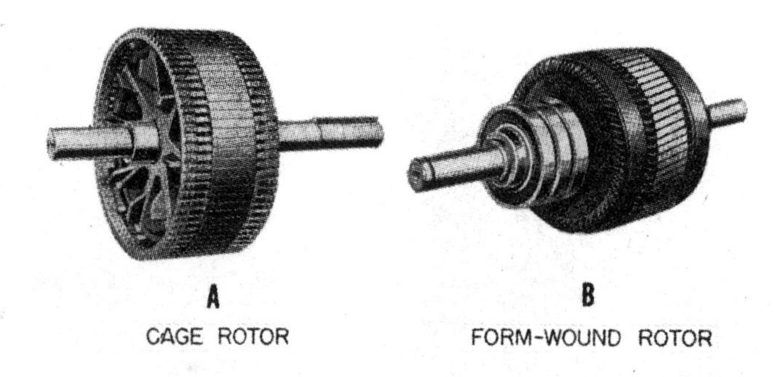

Figure 3–10 Induction motor rotors: (a) squirrel-cage; (b) form-wound. (Courtesy Westinghouse Electric Co.)

power factor, efficiency and pull-out torque somewhat lower than class A.

Class D: high slip: high-resistance squirrel-cage winding; results in high starting torque, low starting current, high slip, low efficiency. Generally used with a flywheel.

Class E: low starting torque; normal starting current: variations of class A. Hardly ever used.

Class F: low starting torque; low starting current: variations of class B. Hardly ever used.

Wound-Rotor Type

In this type of rotor, the conductors are wound around the core and connected in wye or delta (for three-phase), with the terminals brought out to three slip rings mounted on the rotor shaft. The slip rings, connected through brushes, are provided so that the coils of the (three) phases may be short-circuited for normal operating conditions, essentially duplicating the squirrel-cage type (Figure 3–10b). They may also be connected together through suitable external resistances used to increase the starting torque or to vary the speed of the motor. As the voltage induced in the conductors is usually low, the insulation around the conductors need only meet minimum electrical and mechanical standards (Figure 3–11).

The chief advantage of this type of rotor is its ability to insert resistance between its slip rings. This makes possible high starting torque with moderate starting current, smooth acceleration under heavy load, less heating during starting, and adjustable speed. Its disadvantages are its higher initial and maintenance costs, slightly higher resistance, and less ruggedness than the squirrel-cage type.

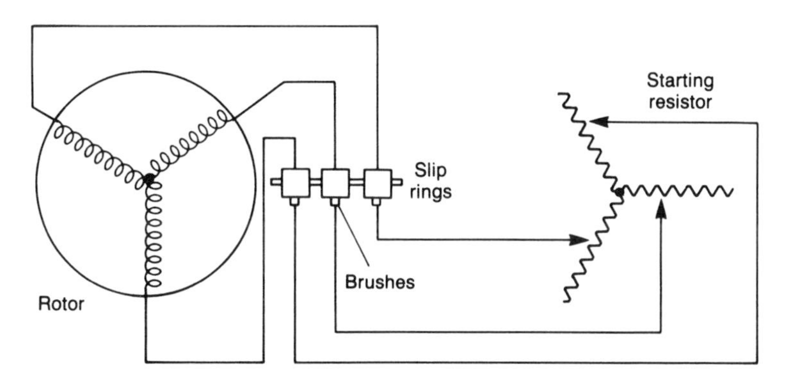

Figure 3–11 Wound rotor with external variable resistance.

THREE-PHASE AND TWO-PHASE INDUCTION MOTORS

All of the discussion has been centered on three-phase motors, but it applies equally well to two-phase motors. The only differences are that there are three coils in the stator field and in the wound-type rotor of the three-phase motor, while two-phase motors have two coils in their stators and wound-type rotors. Here the two-phase voltages are 90° or one half-cycle apart (Figures 3–12 and 3–13). The rotating magnetic field produced in a two-phase induction motor is shown in Figure 3–14. Two-phase motors (and their associated supply systems) are few and rapidly becoming obsolete.

OPERATION AS A SINGLE-PHASE INDUCTION MOTOR

Should one phase of the three-phase (or two-phase) supply become deenergized for any reason, the polyphase motors will not have all of their coils energized. If the stator or rotor coils are connected in wye for a three-phase motor, two

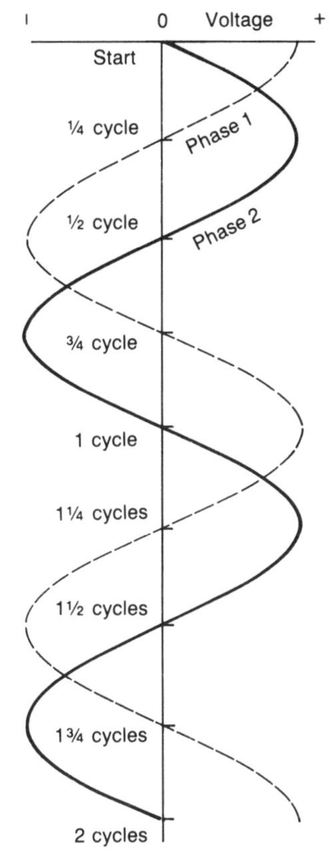

Figure 3–12 Voltage variations: two-phase alternating-current system.

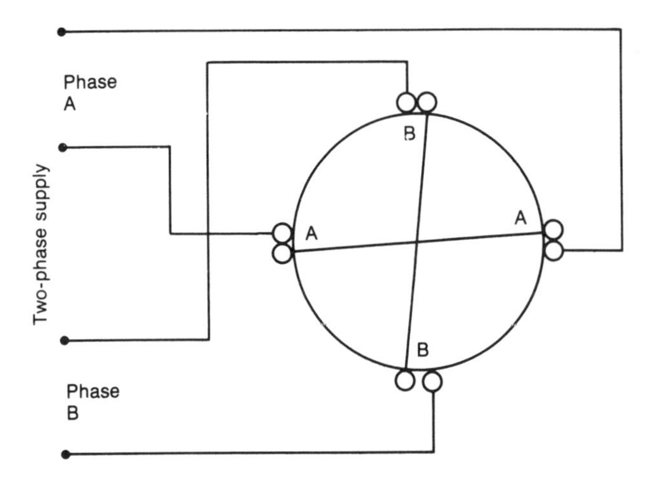

Figure 3–13 Simplified diagram of two-phase stator.

of the coils will remain energized from the remaining two energized phases; if connected in delta, one coil will have full-phase voltage supplied, while the other two will be in series across the two energized phases and each will experience half the line voltage. In either case, the motor will have a single-phase line-to-line voltage across two of its three terminals (Figure 3–15).

If the interruption to one phase should occur while the motor is operating at or very near normal (synchronous) speed and the load does not change, the motor will continue running as a single-phase motor, but the available torque may be lessened. If the load exceeds the available torque, the motor may stall. Once stopped, however, the motor cannot be started again by itself, but will require some means of starting (see Chapter 5). Obviously, too, such a motor cannot be started initially from a standstill condition if the incoming supply has one phase deenergized.

Rotating Field

When operating as a single-phase motor, the stator field will continue to provide a rotating field, but it will not be so smooth and even as when operating three-phase. The rotating magnetic field with one phase deenergized is shown in Figure 3–16. Compare this with that of Figure 3–6; position numbers 3, 5, 9, and 11 will have a monentary interruption of the field, while position numbers 2, 4, 6, 8, 10, and 12 will have a weakened field. The flywheel effect of the heavy metal mass of the rotor will overcome these deficiencies. This may be reflected in a tendency to uneven rotation, some vibration may occur, and the motor may run hotter and noisier. Similar action takes place when a two-phase motor operates with one supply phase deenergized.

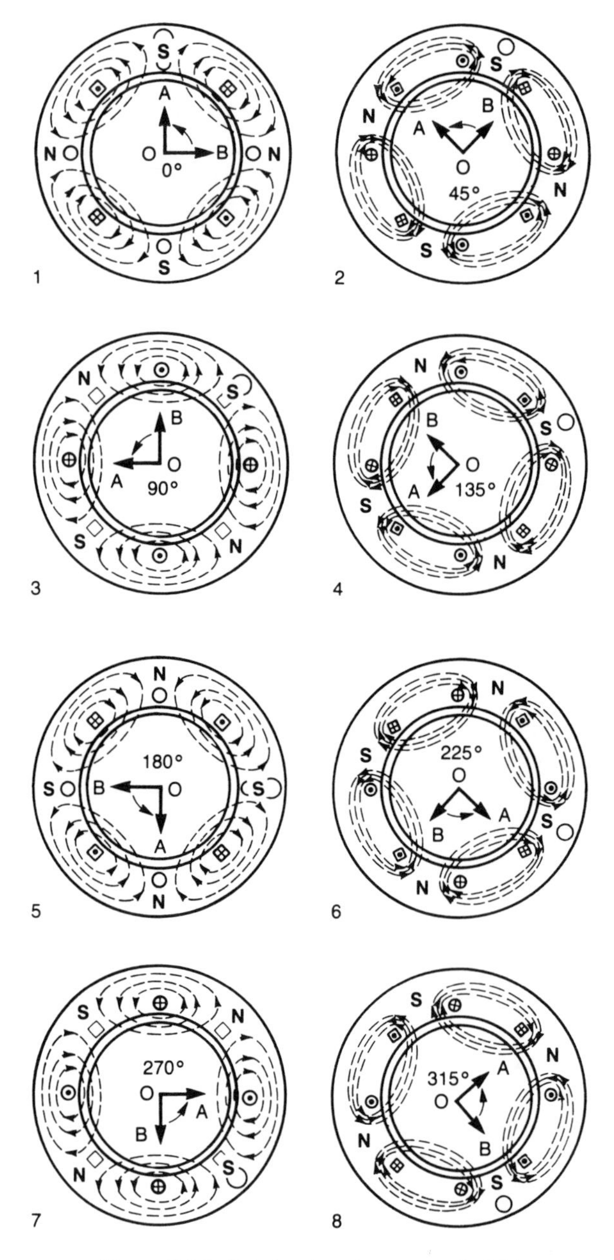

Figure 3–14 Rotating two-phase magnetic field.

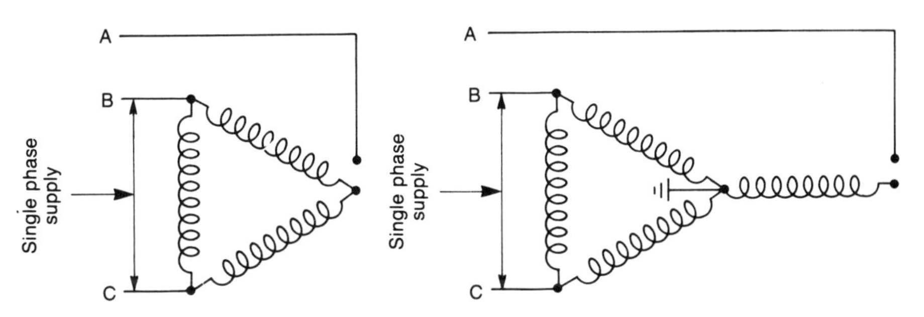

Figure 3–15 Field connections with one phase deenergized.

Locked-Rotor Condition

If the induction motor is stalled, whether from overload, mechanical, or electrical failure, the motor will act as a transformer (see Chapter 10). Each phase of the field (stator) induces a voltage in the rotor, and polyphase currents will flow in the rotor of the same frequency as that of the supply. These currents will reflect the ratio of transformation that exists between the number of turns of the field and those of the rotor windings:

$$\frac{\text{rotor current}}{\text{stator current}} = \frac{\text{stator winding turns}}{\text{rotor winding turns}} = \text{ratio of transformation}$$

The currents in both the stator and rotor windings will be restricted relatively slightly by the resistances of the windings and by the air gap between the stator and rotor. This, in effect, constitutes a break in the "transformer" core, interposing a resistance to the motion of the magnetic lines of force in the core.

The resulting increased rotor current lowers the counter EMF produced in the field coils and causes the field current to tend to increase. This excessive current may damage the motor windings unless protective devices act to disconnect the motor from the source of supply, or act to decrease the supply voltage that will restrict the current flow to safe values.

LOSSES

The losses on an induction motor include copper losses and core losses in both stator and rotor, and friction and windage losses. For practical purposes, core, bearing friction, brush losses (if any), and windage losses may be considered constant for all loads on induction motors having small slip, usually not over 10% (of the synchronous speed).

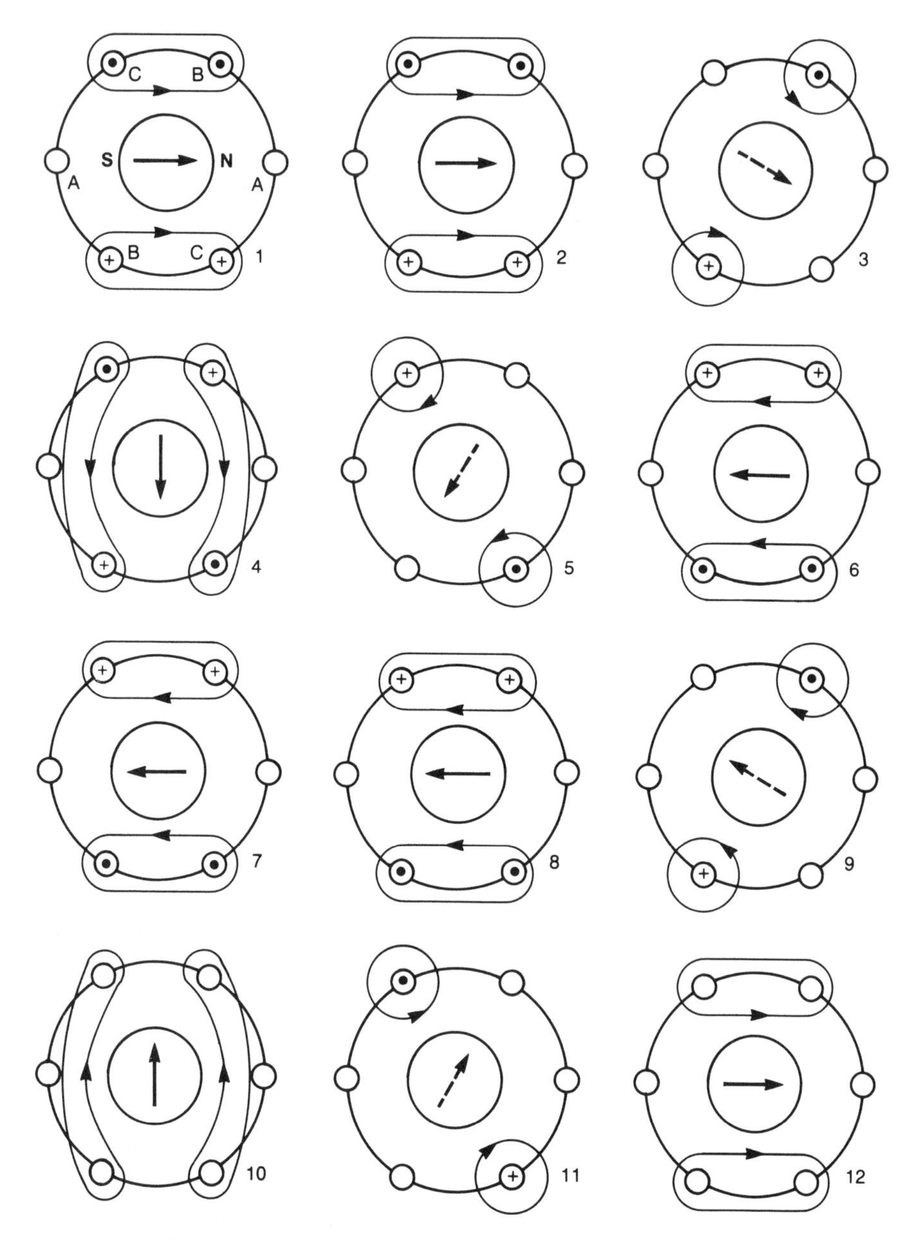

Figure 3–16 Rotating magnetic field with one phase deenergized.

EFFICIENCY

The efficiency of any machine is the ratio of the output power to the input power. For polyphase induction motors of both squirrel-cage and wound-rotor types, at full rated load, it may vary from about 85% for small motors to more than 90% for large motors. Some typical characteristic curves for such motors, showing the variations in efficiency with power factor, slip, torque, and input current, are shown in Figure 3–9b. Actual efficiencies may vary moderately from those shown depending on the variations in construction of the motors to accommodate the purposes for which they are to be employed: for example, high torque, low starting current, for use when starting under load (e.g., compressors); high-slip motors for high intermittent starting (e.g., punch presses).

CONTROL: STARTING AND SPEED CHANGES

To limit the initial inrush of current, when necessary, some type of starter is usually employed. In many instances, the starting device is also designed to provide control of the speed of the motor while operating.

An alternating-current induction motor, under certain conditions, may be started by connecting the stator directly across the main supply line. This may be feasible if the motor is not too large (usually 5 horsepower or less) and the supply system has sufficient capacity to prevent the development of objectionable voltage conditions (such as flicker and voltage fluctuations) at the consumer and in the neighboring area.

One way to lower the inrush current is to reduce the applied voltage. Inserting a resistance in the stator circuit will reduce the inrush current and also the strength of the rotating magnetic field (Figure 3–11). In turn, the current induced in the rotor will also lessen and the interaction between the two magnetic fields will cause the rotor to turn at a slower rate. At the same time, it will generate a counter EMF in the stator field coil which will permit full voltage to be applied when the resistance is short-circuited (taken out of the circuit, which may be done automatically). If the resistance is left in the circuit and is made variable, the rate of speed of the motor can be controlled. If the resistance is used only for intermittent starting, its materials may be subjected to temporary higher temperatures and need not be as rugged as when it is used for continuous operation for speed control.

One other way to achieve the same result is to insert a variable resistance in the rotor circuit, which is possible in the wound-type rotor (Figure 3–17). The same internal interaction and the same construction limits generally will also apply. Since it is simpler to insert the resistance in the stator circuits (no moving parts), this type of control is made rugged for continuous operation and usually used with large motors.

A more common method for reducing and controlling voltage employs

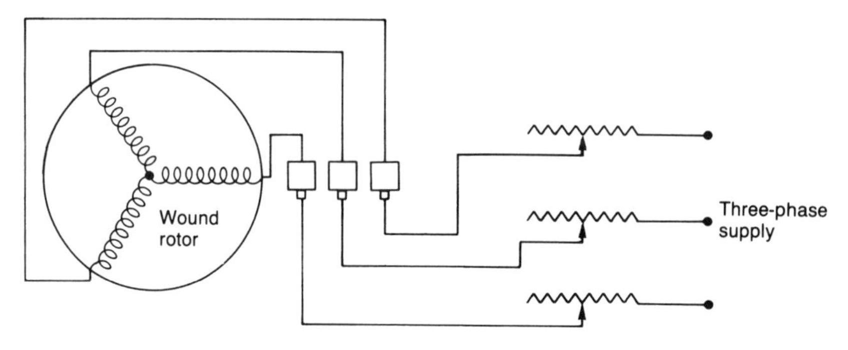

Figure 3–17 Starting arrangement with variable resistance in wound-rotor circuit.

autotransformers, sometimes referred to as a *compensator*. A simplified schematic diagram is shown in Figure 3–18. Taps on the autotransformer windings will apply a lower voltage to the motor. After a proper interval, during which acceleration occurs, full line voltage is applied to the motor. The interval may be based on a fixed period or may be determined by the speed of the motor; the operation may also be accomplished automatically. The autotransformer has the advantage over the resistance in permitting the motor to draw a relatively high starting current from the autotransformer secondary with a relatively low supply-line current (see Chapter 10).

APPLICATIONS

The several types of rotors described above are employed to meet certain applications for which they may be best suited. A few are listed in Tables 3–1 and 3–2, with indications of the nature of service.

Specifications for motors should therefore take into consideration the use to which they are to be employed.

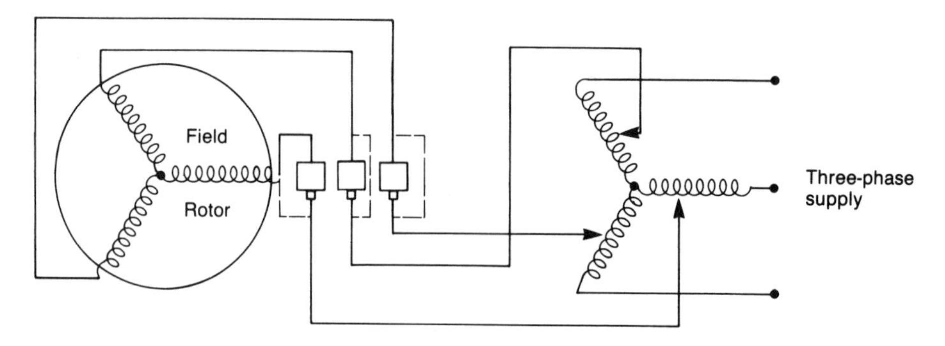

Figure 3–18 Starting arrangement with autotransformer in wound-rotor circuit or field circuit (dashed line).

TABLE 3–1 **Squirrel-Type Rotors**

Low-Resistance Single Cage	High-Resistance Single Cage	Double-Deck Cage
Motor-generator sets	Flywheels: punches, shears	Plunger-type pumps
Centrifugal pumps	Cranes, hoists	Large blowers
Fans and blowers	Elevators	Applications where high starting currents are not permitted but high efficiency is expected
Line-shaft drives	Laundry machines	Applications where danger of fire and explosion preclude use of slip rings and brushes
Cement mixers	Frequency changers	
Grinders	Centrifuges	
Rotating kilns	Brake motors	
Tube and ball mills	Valve motors	
Lathes	Spinning mules	
Shapers		
Paper machinery		
Numerous miscellaneous uses requiring high starting torque and constant speed		

TABLE 3–2 **Wound-Type Rotors**

For Constant Speed	For Variable Speed
Air compressors	Hoists, capstans
Ice machines	Cranes
Flour mills	Elevators
Belt conveyors	Flywheel motor–generator sets
Ship propulsion	Steel mill machinery
Locomotives	Coal and ore loaders and unloaders
Rock and ore crushers	Electric shovels
Paper machinery	
Steel machinery	
Dredging machinery	

REVIEW

- In an alternating-current induction motor, each of the three phases of a three-phase supply are displaced 120 electrical degrees apart (Figure 3–3), and each coil of the stator will produce an alternating magnetic field displaced 120° apart (Figure 3–5). The result is that the combination of the three magnetic fields of the stator will form a rotating field in the air gap revolving at a constant (synchronous) speed depending on the frequency of the alternating-current supply (Figure 3–6).

- The rotating magnetic field of the stator cuts the conductors of the rotor, inducing currents in them that produce a magnetic field similar to that produced in the stator. The interaction of these two magnetic fields develops a torque that causes the rotor to turn (Figure 3–7).

- There is no external electrical supply to the rotor conductors; the currents in the rotor are entirely induced. The rotor conductors are connected together at each end of the rotor, completing their circuits (Figure 3–8).

- If the rotor could turn at synchronous speed, there would be no relative motion between the conductors of the rotor and the rotating stator field and therefore no current induced in the rotor conductors. With no rotor magnetic field to interact with the rotating stator magnetic field, no torque would be produced.

- Because of losses in the rotor and load imposed on it, the rotor will rotate at a less-than-synchronous speed. This will cause a relative motion between the rotating stator magnetic field and the rotor conductors, inducing in them a current. In turn, this produces a rotor magnetic field that will interact with the rotating stator magnetic field producing a torque necessary to overcome the resistance to turning caused by the rotor losses and load.

- The difference between the speed (synchronous) of the rotating stator magnetic field and the resultant speed of the rotor is known as the slip of the induction motor, and is expressed in percent of the synchronous speed. In practice, it is normally about 10% or less.

- The torque produced in the rotor will be directly proportional to the current in the rotor and the strength of the stator magnetic field. It increases as the slip increases up to a point where further slowing of the rotor can no longer be accompanied by strengthening of the stator (rotating) field because of the magnetic saturation of the stator core. This point is called the pull-out point and the motor stalls.

- There are two distinct types of rotors: the squirrel-cage type and the wound-rotor type (Figure 3–10). As its name implies, the squirrel-cage type consists of conductors placed longitudinally on the rotor core and connected together at both ends; in some instances, there may be two layers of such conductors on the rotor surface.

- The second or wound-rotor type has conductors on the rotor surface but brought out to slip rings and brushes. This is done to enable a resistance to be placed in the circuit to reduce the starting current where that current may be so large as to affect the supply system; as the rotor reaches normal speed, the slip rings are automatically short-circuited, in effect making the wound-rotor type a squirrel-cage type of rotor.

- ✦ The same actions apply to two-phase motors as apply to three-phase motors (Figure 3–14).
- ✦ Should one phase of the supply to a polyphase motor become deenergized, the motor will continue to operate as a single-phase motor but at lowered efficiency. If the motor is stopped for any reason, it cannot start again until polyphase supply is restored (Figures 3–15 and 3–16).
- ✦ Control of starting current as well as of the motor speed may be obtained by inserting a variable resistance in the stator circuit of a squirrel-cage motor or in the rotor circuit of a wound-rotor motor. An autotransformer with taps may be used in place of the variable resistance (Figures 3–17 and 3–18).
- ✦ Efficiencies of induction motors may vary somewhat depending on its characteristics: high or low starting torque; high or low starting currents. Nevertheless, the efficiency of the polyphase induction motor ranges from about 85% to over 90%.

STUDY QUESTIONS

1. What is the rotating magnetic field of the stator of an induction motor? How is it produced?
2. What two factors determine the speed of the revolving field, and what is the speed called?
3. How may the direction of rotation of the rotating magnetic field be changed in a three-phase winding?
4. What is a squirrel-cage rotor, and how is current supplied to it?
5. What is a wound rotor, and how is current supplied to it?
6. How is the torque of an induction motor affected by an increase in the rotor current? In the stator field strength?
7. What is meant by saturation of a magnetic core? What happens when the stator core is saturated and more load is imposed on the induction motor?
8. What is the pull-out point? What happens to the motor at this point?
9. How may starting and running currents of an induction motor be controlled?
10. What are three types of losses in an induction motor?

chapter 4

Alternating-Current Synchronous Motors

INTRODUCTION

The idea of an alternating-current synchronous motor was almost immediately apparent with the development of alternating-current generators. Early motors were of the rotating armature type, similar to the generators but equipped with commutators or other starting devices (such as an induction motor). The principle of the rotating magnetic field propounded by Ferraris in 1885 was instrumental in the development not only of the induction motor, but of the synchronous motor as well. Original applications of synchronous motors were limited to easy starting conditions. Improvements in its features, coupled with the advantages of power factor correction, considerably widened its employment.

PRINCIPLE OF OPERATION

Like all other motor types, the driving torque of an alternating-current motor is produced by the interaction of magnetic fields set up in the stator and rotor of the motor. In the synchronous motor, the supply polyphase alternating current to the stator produces an alternating magnetic field which, as in the induction motor, is a rotating magnetic field (see Figure 3–6). Unlike the induction motor, however, a direct-current source is supplied to the rotor, producing a magnetic field of fixed polarity at each pole.

When the stator and rotor are energized, a "north" pole produced for half a cycle by the rotating stator magnetic field interacts with the north pole produced in the rotor, the opposing force set up tends to cause the rotor to turn in one direction. In the next half-cycle, the "south" pole produced in the stator and the attracting force set up tends to cause the rotor to turn in the opposite direction. The net result is that the torques thus produced cancel each other; that is, no torque is produced, and the motor does not turn (Figure 4–1).

If, however, the rotor is somehow caused to rotate at the synchronous speed, the speed of the rotating stator magnetic field, then the torque set up by the "north" pole of the stator field interacting with the north (or south) pole of the rotor will continually interact with each other; this causes the rotor to turn at the same speed as the rotating stator magnetic field, that is, at synchronous speed.

Starting

There are two ways in which the synchronous motor may be brought up to synchronous speed before load is applied. A nonsynchronous motor, that is, an alternating-current induction motor or a direct-current motor, can be used for this purpose. The induction motor is not usually employed. A direct-current motor coupled to the rotor shaft may be used to bring the rotor up to synchronous speed, at which time the direct-current motor can be operated as a direct-current generator to supply the direct current to the rotor.

More often, however, a squirrel-cage winding (sometimes known as a *damper winding*) is incorporated in the rotor of the synchronous motor; the direct-current field is temporarily deenergized and a reduced polyphase alter-

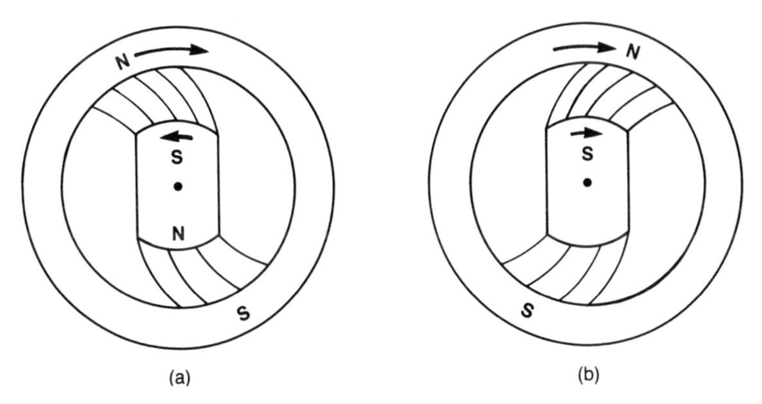

Figure 4–1 Action of synchronous motor when first energized, showing tendency of rotor to turn counterclockwise (a) or clockwise (b).

nating-current voltage supplied to the stator. As soon as the rotor reaches synchronous speed, the full voltage is applied to the stator field winding and the rotor energized from a direct-current source through a resistance to hold the line current to a small value.

If the polarity of the field produced instantaneously in the stator is incorrect in relation to that of the rotor, the counter EMF produced in the stator windings will cause the current flowing to increase, causing the rotor to slow momentarily until the next pole (of the opposite polarity) of the rotor will bear the correct relation between the two fields. The two will then be "in step"; the slowing until the polarity of both the rotating stator field and that of the rotor are correct is called *slipping a pole*.

Caution must be observed at startup when the direct-current supply to the rotor is disconnected. As the stator winding is energized from a polyphase alternating-current source, the rotating stator magnetic field will cut the windings of the rotor, inducing in it a high alternating-current voltage. Connecting a low resistance across the rotor direct-current winding will permit an alternating current to flow through the direct-current winding, creating a counter EMF that will restrain the induced voltage to a safe value.

Starting Torque

The interaction of both magnetic fields produced in the rotor (the squirrel-cage damper, the direct-current winding) and the rotating stator magnetic field produce the torque that causes the motor to turn. It must be remembered that in an induction motor the torque is produced by a difference between the speed of the rotor and that of the rotating stator magnetic field; the difference is called the slip. In the synchronous motor, there is no difference in speed, that is, no

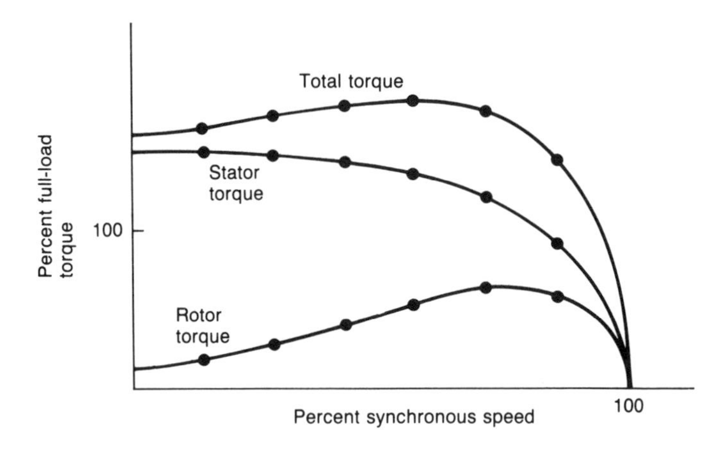

Figure 4–2 Starting torque of a synchronous motor.

slip between the rotating stator magnetic field and the rotor (except momentarily, as detailed later, as load is applied).

The torques produced by the two windings at different speeds during the starting period are shown in Figure 4–2. Note that the torque produced by the current in the rotor coil helps sustain the starting torque as the motor approaches synchronous speed, and that both windings produce no torque at synchronous speed since there will be no relative motion between the fields of the rotor and stator, and the induced voltage is zero, while no direct current is yet supplied to the rotor direct-current winding.

Running Torque

As load is applied to the motor, the rotor will momentarily slow down, causing the relationship between the pole positions of the stator and rotor magnetic fields to be displaced; the pole of the rotor will lag behind the center of the poles formed by the rotating stator magnetic field (Figure 4–3). This difference in position, called the *displacement* of the rotor, causes the interaction between the fields to change, reducing the counter EMF in the stator winding, causing the "load" current to increase. The increased current will, in turn, strengthen the rotating field and restore the speed of the rotor to its synchronizing value.

Although the synchronous speed of the rotor is restored, the relative physical position of the magnetic poles of the rotor will remain displaced by a small angle. The interaction between the two magnetic fields, displaced by that angle, will in turn cause the counter EMF in the stator to combine with the applied line voltage in such a way as to cause the resultant voltage and current

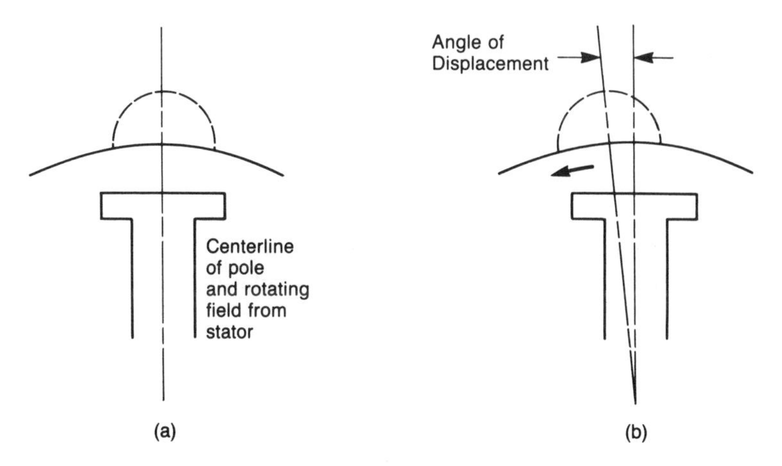

Figure 4–3 Angular displacement of pole and rotating field: (a) pole in phase with rotating field; (b) pole lagging behind rotating field.

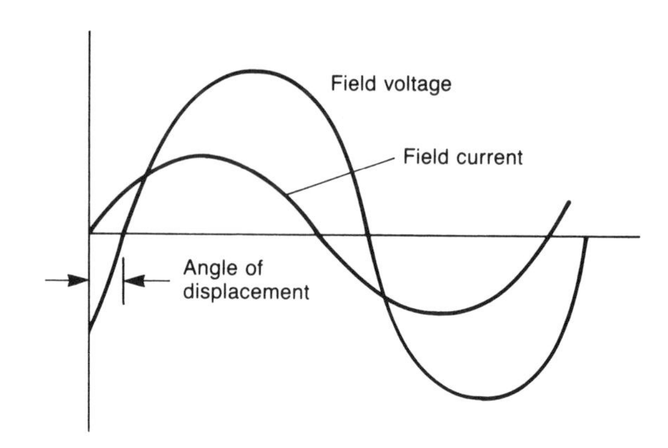

Figure 4–4 Field current lagging voltage by angle of displacement.

in the stator to be displaced from each other; that is, the line current will lag behind the line voltage (Figure 4–4).

The two quantities, voltage and current, however, can be brought to work completely together again, that is, to be in phase again. This may be done by deliberately strengthening the rotor magnetic field (by increasing its direct-current supply) as the load is increased, causing the rotor to speed up temporarily until the poles of the stator and rotor will once more come together and the current and voltage of the supply line will be back in phase.

If the rotor field direct-current supply is further increased, the rotor will again speed up temporarily so that the relation between the stator and rotor magnetic fields will be such that the current can be made to lead the supply line voltage.

Thus, for a definite load, the relation between the current and line supply voltage may be controlled by varying the field (rotor) direct-current supply. A weak field can cause the current to lag the supply-line voltage, and a strong field can cause the current to lead the supply-line voltage. Normal field current for a given load occurs when the line current and voltage are in phase; that is, the power factor is unity or 100%.

The variations in line (armature) current for different loads and field current and the corresponding values of power factor are shown in Figure 4–5; these are sometimes called the synchronous motor V curves.

This phenomenon may be illustrated by picturing two people running together side by side, in synchronism, each carrying one end of a rod (representing a load); one person may represent the stator and the other the rotor. If a load is imposed on the rotor, that person will tend to slow down (lose speed), lagging more and more behind. However, by applying greater effort and expending more energy, that person can be made to regain and maintain his (or her) original speed, but still remain (lag behind) a short distance from his (or her) partner. If more effort is applied to the rotor person, that person can be

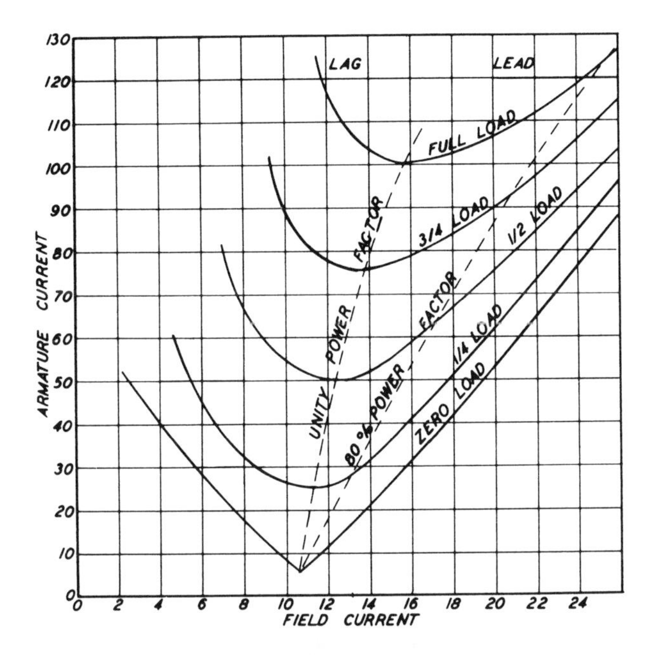

Figure 4–5 V-curves showing relationships between armature current and field current at different loads. (Courtesy Westinghouse Electric Co.)

made to speed up until both persons are once again running side by side (in synchronism and in phase). If still more effort is applied to the rotor person, that person can be made to run ahead of (or lead) his partner, each running at the same speed (in synchronism) but displaced a distance from each other (Figure 4–6).

It may also be observed that when the two persons are running side by side (in phase), their entire efforts are used in moving the load in a straight line. When one person is lagging the other, there will be a tendency for them to run at an angle away from a straight line toward the lagging person; only a portion of their effort will move the load in the straight line, and a portion will be expended to overcome the tendency to move in the direction indicated in the figure. When that person is leading the other, a similar action takes place, but the tendency to run in a direction other than the straight line will be in the opposite direction, as shown in Figure 4–6.

OPERATION AS A SYNCHRONOUS CONDENSER

This is an important phenomenon, as synchronous motors connected to a supply line can influence the power factor of the power flowing in that line. The nearer to 100% power factor, the greater the amount of effective power can be delivered

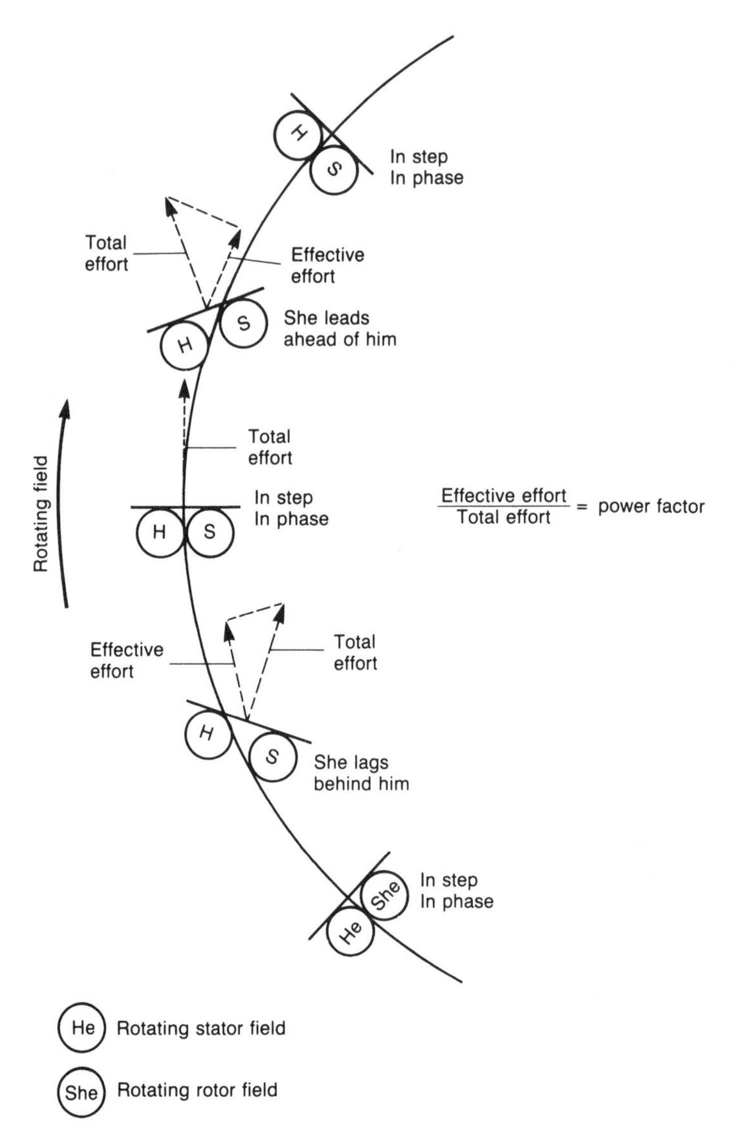

Figure 4–6 Synchronous motor analogy. *Note*: Except for lag and lead instants, both "he" and "she" run at the same (synchronous) speed.

with the same current flowing in the supply line. This action is somewhat similar to the action of a capacitor (see Chapter 10).

A capacitor, or condenser, operates to oppose the free flow of an alternating current in a conductor similar to, but opposite in effect, as a magnetic field. When two conductors are near each other, the alternating magnetic field of one will induce a voltage in the other, opposing that which may exist in it; it is sometimes referred to as a counter electromotive force, or counter EMF. Its effect is to have the current in the second conductor to lag its voltage. A capacitor achieves somewhat the same effect electrostatically, except that the opposition current created in the adjacent conductor will be in the opposite direction to that created electromagnetically. For a more complete explanation, see Chapter 10.

If no load is applied to such a motor, its field current may be adjusted to have it operate at a very low leading power factor, of about 10%; it is generally referred to as a *synchronous condenser*. Only enough (in phase) power is required to supply the losses in the motor, but enough current at a *leading* power factor will be drawn from the supply line as to cancel current at a *lagging* power factor that may be drawn by other loads connected to the same supply line, thereby improving the supply line power factor.

OPERATION AS A SINGLE-PHASE MOTOR

Should one phase of the three-phase supply fail, the motor will continue to run as a single-phase induction motor, similar to the operation of a three-phase induction motor described in Chapter 3. Similarly, a three-phase synchronous motor cannot be started when one phase is deenergized, nor can it be restarted once it comes to a stop.

EFFICIENCY

The efficiency of a synchronous motor, like that of any piece of equipment, is the ratio of its output to its input; the input is equal to the output plus its losses. The losses are similar to those found in other motors: resistance losses in both field and armature, called copper losses; stray current losses in the cores of both field and armature, called iron losses; and friction and windage losses. In general, the losses in a synchronous motor, compared to those in an induction motor, are lower principally because of operation at a better power factor; hence their efficiencies are generally better than those of induction motors. In general, efficiencies for synchronous motors are in the range 90 to 95%.

The importance of efficiency depends on the type of load and on the amount of time it will operate. If it is to run continuously at full load for months (e.g., municipal water works), a very small difference in efficiency may translate into substantial power cost. If used only intermittently (e.g., pumping water during severe floods), a few hours in one or more years, the efficiency is of little or no importance. In general, the efficiency should reflect the maximum possible losses without exceeding temperature limitations.

CONTROL: STARTING AND SPEED CHANGES

Described earlier are methods of starting synchronous motors: turning of the rotor by means of a second motor (direct-current or induction motors), accelerating it until it reaches synchronous speed; turning of the rotor by means of damper windings in the rotor, essentially operating as an induction motor, while the direct-current field is disconnected, until synchronous speed is reached, at which point the direct-current field is manually or automatically energized. Like the induction motor, starting may be accomplished by connecting it directly across the line if the motor is of sufficiently small size as not to affect the line to which it is connected. Starting at a lower applied voltage may also be accomplished by the insertion of a resistance or using an autotransformer, as described for induction motors.

Since this type of motor turns only at a constant speed, speed control as such is not needed. As load is applied to the motor, the speed will tend to decrease and adjustment of the direct-current field current will be necessary to prevent the speed decreasing to the point where the motor will stall at the pull-out point described earlier. Speed–torque characteristics of the damper winding with three different resistances are shown in Figure 4–7: curve 1 has a high resistance, curve 2 a medium resistance, and curve 3 a low resistance; the lower

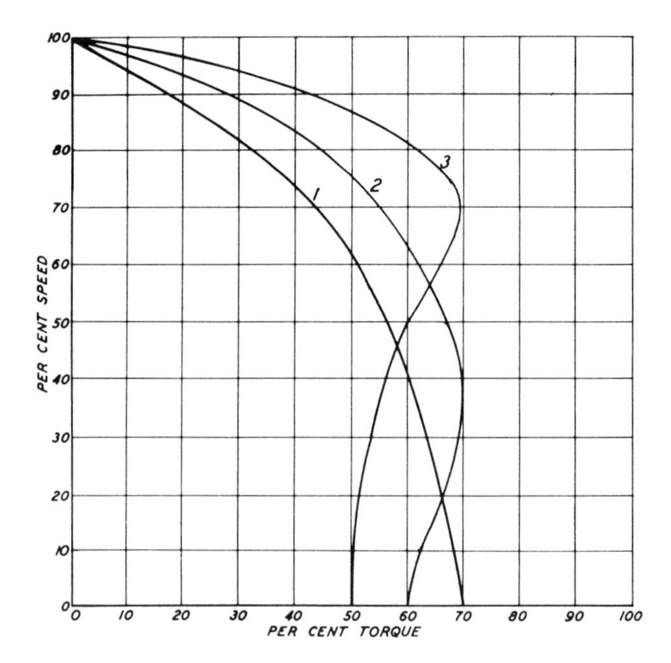

Figure 4–7 Speed–torque characteristics of damper winding: Curve 1, high resistance; curve 2, medium resistance; curve 3, low resistance. (Courtesy Westinghouse Electric Co.)

the resistance, the higher the armature current and the higher the maximum torque.

The other major control, that of improving power factor by adjusting the direct-current flow to the rotor, has been described previously.

APPLICATIONS

Applications for synchronous motors may be divided into three general classifications:

1. *Machines where partial unloading during starting is practical*: centrifugal fans and pumps; compressors; rubber mixers, motor–generator sets; locomotives; ship propulsion
2. *Machines where starting loads consist of friction and inertia only*: reciprocating compressors; grinders; paper mills; centrifugal pumps; centrifugal fans.
3. *Machines starting essentially under load*: cement mixers; rubber mixers; crushers; band saws; line shafts; metal and steel mills; large-draft fans; ship propulsion

REVIEW

✦ The stator of a polyphase synchronous motor produces a rotating magnetic field that revolves at synchronous speed similar to that of the induction motor. The rotor coils, however, are supplied from a direct-current supply that produces a magnetic field of fixed polarity at each pole.

✦ When both stator and rotor are energized at standstill, the stator rotating alternating magnetic field and the direct-current nonalternating magnetic rotor field interact in such a way that no torque is produced and the rotor will not turn (Figure 4–1).

✦ If the rotor can be made to turn at synchronous speed, the same speed as the stator rotating magnetic field, the two will interact and the rotor will turn at the same speed as the stator rotating magnetic field, that is, at synchronous speed.

✦ To start a synchronous motor, the rotor may be brought up to synchronous speed by some means. An external induction or direct current motor may be used. More commonly, "damper" windings are installed in the rotor that cause it to act as a squirrel-cage induction motor with the direct-current supply temporarily disconnected. When the alternating field set up in the rotor interacts with the stator rotating magnetic field, the two fields will be "in step" and the direct-current supply to the rotor reenergized.

- ✦ In the synchronous motor, there is no difference in speed between the rotor magnetic field and the stator rotating magnetic field; that is, there is no slip as in the induction motor.
- ✦ As load is applied, the rotor will slow down momentarily and the relationship between the stator and rotor magnetic fields is displaced a small amount. The difference in position, the displacement, causes the fields to interact, causing the current in the stator to increase with the load. The increased current strengthens the rotating field and restores the speed of the rotor to its synchronizing value (Figure 4–3).
- ✦ Although the synchronous speed of the rotor is restored, the relative physical position of the magnetic poles of the rotor and stator will remain displaced by a small angle. The new relation between the rotor field and the stator rotating field causes the current to lag behind the supply voltage (Figure 4–3).
- ✦ The two fields can be brought back into step at synchronous speed by increasing the strength of the rotor field by increasing its direct-current supply. The poles of the rotor and stator will once again come together and the supply alternating current will be back in phase with the supply voltage (Figure 4–6).
- ✦ If the direct-current supply to the rotor is increased further, the rotor will speed up further so that the relation between the stator and rotor fields will be such that the supply current will lead the supply voltage.
- ✦ This phenomenon makes possible the use of synchronous motors to cause the supply current to lead (or lag) the supply voltage, acting in the same manner as a capacitor connected to the supply line; hence it is sometimes called a synchronous condenser.
- ✦ Like an induction motor, the synchronous motor can continue to run when one phase of a three-phase (or two-phase) supply is deenergized. Once it stops, however, it cannot be restarted until all three phases are energized.
- ✦ Since the synchronous motor is made to start as an induction motor, the same starting controls are used to bring the motor up to speed as are used for induction motors.

STUDY QUESTIONS

1. What is synchronous speed?
2. What is the difference in the relationship of the rotor field to the stator rotating field for the synchronous motor as compared to the induction motor?

3. How is current supplied to the conductors of the rotor? How does this differ from that of an induction motor?
4. What is a damper winding in a synchronous motor, and what is its function?
5. Why is a low resistance connected across the direct-current field winding of a synchronous motor during the starting period?
6. For a given load on a synchronous motor, how can its power factor be changed?
7. What is the name commonly applied to the synchronous motor that is used to change the relationship between the current and the voltage of the supply circuit?
8. How will a polyphase synchronous motor operate when one phase of its supply is deenergized?
9. As load is applied to a synchronous motor, what must be done to have it maintain its synchronous speed?
10. What are the losses in a synchronous motor, and how do they compare with those of an induction motor?

chapter 5

Single-Phase Alternating-Current Motors

INTRODUCTION

Single-phase alternating-current motors are a type of induction motor which, as the name implies, operates on one phase. One obvious advantage is that they eliminate the need for a three-phase supply line, requiring only a simple, less expensive, two-wire supply. On the other hand, the relatively lower power capability compared to polyphase motors restricts their use generally to small units. It is most popularly used in fractional-horsepower sizes for appliances found in households and offices, and for small tools and machines. In exceptional cases, especially where polyphase supply may not be readily available (such as for trolley and railroad traction), such motors may range as high as several hundred horsepower.

PRINCIPLE OF OPERATION

A single-phase induction motor is somewhat similar to a polyphase squirrel-cage motor, except that it has only one stator winding. Having no magnetic rotating field at start, it will have no starting torque. If brought up to speed by some external means, however, a current will be induced in the rotor that will produce a magnetic field. This will interact with the stator magnetic field, producing a rotating magnetic field. This, in turn, will cause the rotor to continue to turn in the direction in which it was started.

The problem then reduces to finding methods of producing out-of-phase magnetic fields in the rotor that will interact with the stator magnetic field to produce the starting torque necessary for the rotor to turn. Several methods accomplish this, the name of each identifying the type of motor: split-phase motor; capacitor motor; shaded-pole motor; repulsion start motor; and alternating-current series motor.

Split-Phase Motor

The split-phase motor has a stator composed of two windings, an auxiliary or starting winding and a main or running winding, whose axes are displaced electrically 90 degrees. The starting winding has fewer turns of smaller-size wire than those of the running winding. The two windings are connected in parallel across the single-phase supply line (Figure 5–1a), but because of the different characteristics of the windings, the currents flowing in them will be of different magnitudes and will be displaced from each other as well as from the supply line voltage.

When energized, these two windings will produce a rotating magnetic field that will rotate around the stator at synchronous speed. The rotating magnetic field cuts the conductors of the rotor and induces a voltage and current in them which, in turn, produces a magnetic field. The interaction of the magnetic fields of the rotor and stator causes the rotor to rotate in the direction in which the stator field is rotating (Figure 5–1b). As soon as the rotor begins to turn, the rotor voltage and current are reduced, producing a rapidly decreasing counter EMF in the stator and causing the rotor to accelerate.

As the speed of the rotor approaches synchronous speed, the starting winding is automatically disconnected from the line supply and the motor continues to run on the main winding alone. The rotating magnetic field is then maintained by the interaction of the magnetic fields of the rotor and stator, and the speed of the rotor is, therefore, also maintained.

When these motors are designed to operate at a base voltage of (say) 120 volts, the stator coils are divided into two equal groups connected in parallel. When designed to operate at double that voltage, 240 volts, the groups are connected in series. The starting torque may be as much as 200% of full-load torque and the starting current may be as much as 10 times the full-load current. The direction of rotation may be reversed by interchanging the starting winding leads.

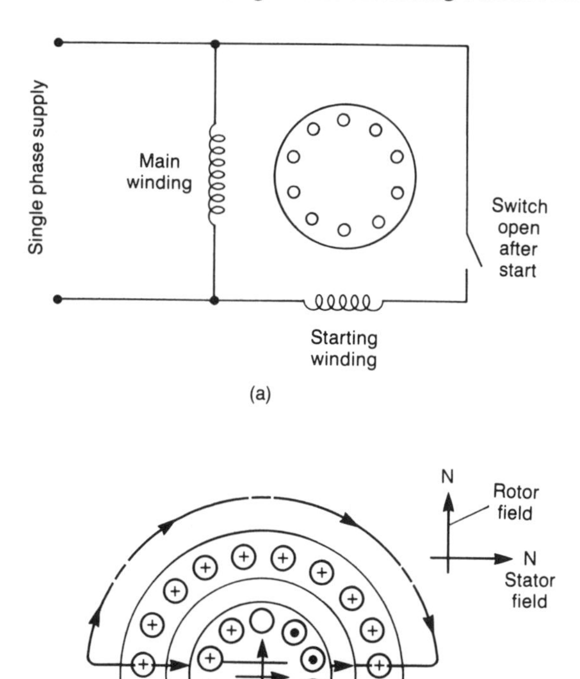

Figure 5–1 Split-phase single-phase motor: (a) circit diagram; (b) operating magnetic fields.

Capacitor Motor

The capacitor motor is essentially an enhanced version of the split-phase motor. A capacitor is placed in series with the starting winding, producing a greater phase displacement of currents in the starting and running windings than is produced in the split-phase motor (Figure 5–2). The starting winding in this case has many more turns of larger-size wire (than for the split-phase motor). The greater phase displacement of the rotor and stator magnetic fields produces a greater starting torque (than the split-phase motor), which may be as much as four times the full-load torque.

With the capacitor in the starting winding and with the motor running at rated speed, the current flowing in the circuit will be less (than for the split-phase motor), thus the need for the winding circuit to be deenergized; in this

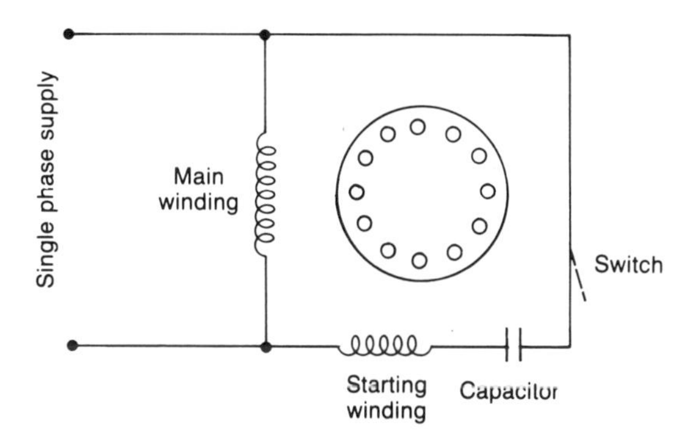

Figure 5–2 Capacitor motor (with switch in circuit, capacitor starts motor.)

case, the motor is referred to as a *capacitor motor*. If, however, it is deemed desirable to take the starting winding and the capacitor out of the circuit after the motor has attained rated speed, the motor is referred to as a *capacitor-start motor*.

Capacitor motors may range in size from fractional horsepower to about 10 horsepower. Like the split-phase motor, the direction of rotation of the capacitor motor may be reversed by interchanging the starting winding leads.

Shaded-Pole Motor

The shaded-pole motor employs a squirrel-cage rotor and a stator with modified pole pieces that accommodate a short-circuited winding (often a copper strap) called a *shading coil*. The arrangement of a four-pole motor of this type is indicated in Figure 5–3. The four coils of the main winding are connected in series across the motor terminals.

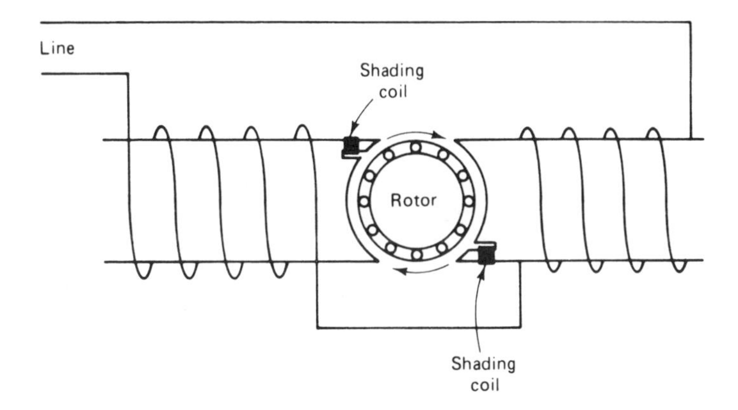

Figure 5–3 Shaded-pole motor.

As the magnetic field of the main pole is increasing, the shading coil is cut by the magnetic field and the voltage and current thus induced in it create a magnetic field that tends to oppose the change in the magnetic field produced by the main stator winding in that portion of the pole it surrounds. The result is that the greater part of the magnetic field rises in the unshaded portions before it reaches its maximum value in the shaded portions of the field.

When the magnetic field reaches its maximum value, its motion is momentarily stopped and the voltage and current induced in the shading coil is zero, and the magnetic field is distributed more evenly over the entire pole face. Then as the magnetic field begins to decline from its maximum value, the induced voltage and current in the shading coil reverse and the magnetic field it produces also reverses its direction, preventing the magnetic field of the unshaded part of the pole from collapsing as quickly as in the region of the shading coil. The effect is a progressive shift in the field from the unshaded to the shaded portions of the poles. The conductors of the squirrel-cage rotor are cut by this moving magnetic field; the interaction between the magnetic field cut by the rotor and that of the moving field of the stator thus created causes the rotor to turn in the direction of the progressively moving field.

The shading coils are left in the circuit after the motor has reached its rated speed, as the loss in them is small and of not much importance in the small motors to which the starting method is limited. As the shading coil operates on only one edge of the pole, the direction of rotation of the motor is not reversible.

The operating characteristics of the shaded-pole motor are similar to those of the split-phase motor. Its simple construction, with no moving electrical contacts, makes it inexpensive and reliable in operation. Its starting torque, however, is low as its efficiency. It may also have a higher noise level than that of other types of single-phase motors.

Repulsion-Start Motor

The repulsion-start motor has a wound rotor with commutator and brushes, and a stator with a distributed single-phase winding. The starting torque is developed through the interaction of the magnetic fields of the rotor and stator. Unlike the split-phase motor, the stator field does not rotate at the start, but alternates. The function of the commutator and brushes is to divide the windings of the rotor into two parallel circuits, as shown in Figure 5–4. Rotor currents are induced by transformer action, and with the brushes as shown in Figure 5–4, create a magnetic field which completely opposes that of the stator, and the result is zero torque. If the position of the brushes is moved as shown, the relative position of the rotor and stator magnetic fields will be such as to interact, causing the rotor to turn in the direction indicated by the position in which the brushes are displaced. As the motor approaches normal (synchronous) speed, a centrifugal device automatically removes the brushes from the commutator

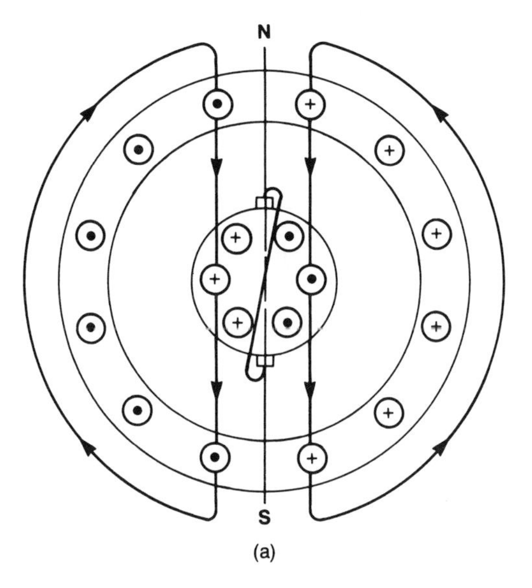

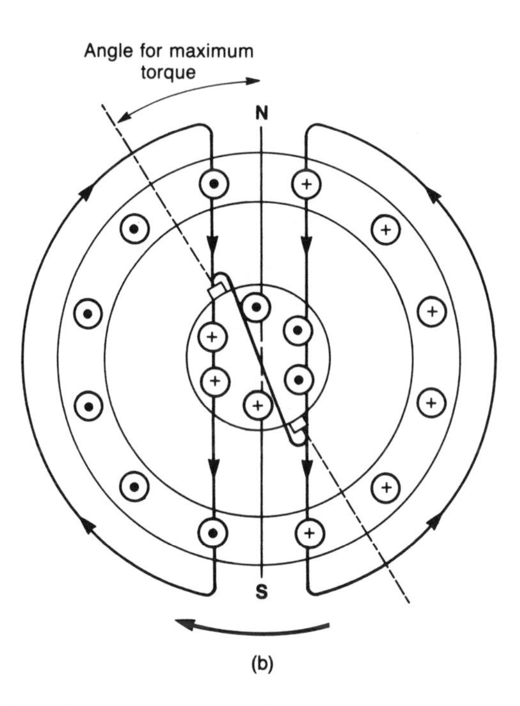

Figure 5–4 Repulsion-start motor: (a) maximum rotor current, no torque; (b) running rotor current, maximum torque.

and places a short-circuiting ring around the commutator; the motor then acts as a squirrel-cage induction motor. The motor derives its name from the repulsion of like poles between the rotor and stator magnetic fields.

The starting torque may be as high as five times full-load torque and the starting current may be as much as four times full-load current. The repulsion-start motor has a higher pull-out torque (the torque at which the motor stalls) than the capacitor start motor. The capacitor-start motor, however, can bring up to full speed loads that the repulsion-start motor can start but cannot accelerate.

Alternating-Current Series Motor

An alternating-current series motor will also operate on direct current. Since the current in the armature and in the field of the series motor reverses at the same time, torque will develop with either alternating- or direct-current input. The alternating magnetic fields of both the rotor and stator in an alternating-current motor, however, produce certain negative effects not found in a direct-

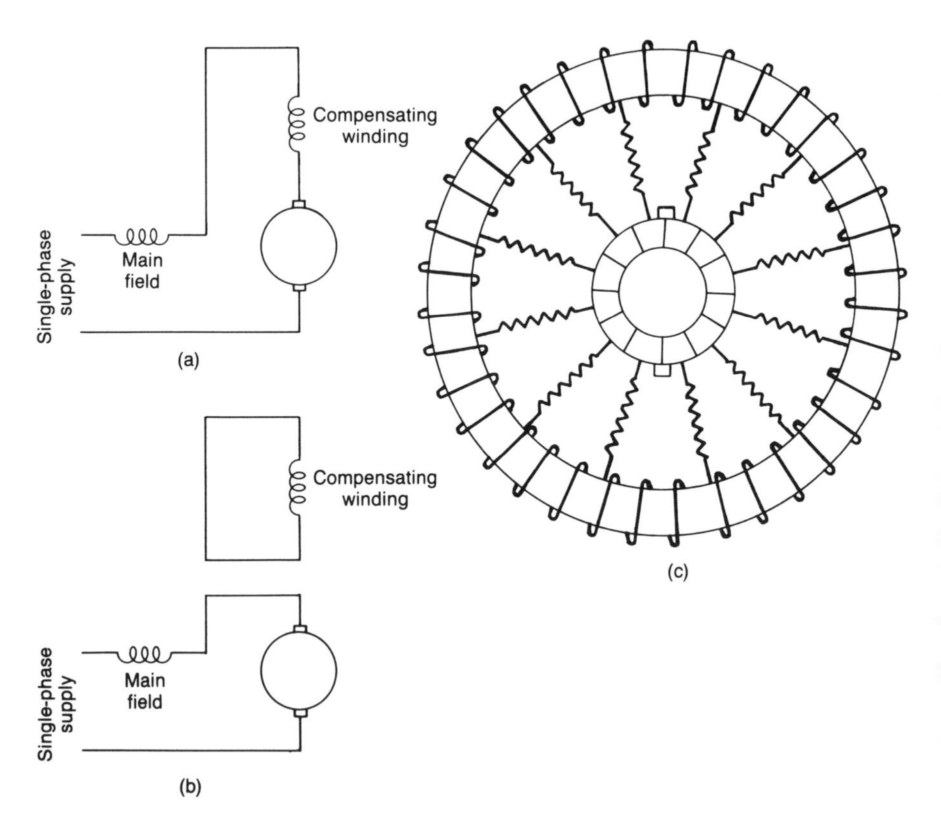

Figure 5–5 Series motor: (a) conductively compensated; (b) inductively compensated; (c) resistances from rotor coils to commutator.

current motor. The inductive effect of the two alternating magnetic fields of the rotor and stator in each other will act to distort these fields, reducing the power developed, resulting in a low power factor, large iron losses in both stator and rotor, and destructive sparking because of the transformer action in the armature coils short-circuited by the brushes during commutation.

To overcome (in part) the effect of the magnetic fields of the rotor and stator on each other, a compensating winding embedded in the pole pieces may be connected in series with the armature (Figure 5–5a); the magnetic field produced by this compensator acts to neutralize the effect of the armature magnetic field, preventing distortion of the main magnetic field. A similar action takes place if the compensator winding is short-circuited on itself (Figure 5–5b). Iron losses are held down by the use of high-permeability steel laminations, while sparking at the commutator is reduced by the use of resistance leads connecting the armature conductors to the commutator segments (Figure 5–5c).

The operating characteristics of the alternating-current series motor are similar to those of the direct-current series motor. If load is removed from the motor, its speed will tend to increase to very high levels that may cause damage or destruction of the motor.

Universal Motors

Fractional-horsepower series motors capable of operating on either alternating- or direct-current are called *universal motors*. They do not have compensators or high-resistance armature-commutator leads.

LOSSES AND EFFICIENCY

The losses of a single-phase induction motor are greater than those of a polyphase motor of the same speed and rating. In the polyphase motor, a three-phase motor, for example, the inductive effects of the three magnetic fields tend to cancel each other out, while in the single-phase motor, the effects of the rotor and stator fields are largely unopposed. Further, in the three-phase motor, each phase takes about one-third of the total current necessary for its rated load, while the single phase of a single-phase motor supplies the entire current. Since the losses vary as the square of the current (I^2R), the sum of the squares of the three small currents is decidedly less than the square of the total of the three small currents: for example, if $I = 3$ amperes, then

$$3(\text{phases}) \times (\tfrac{1}{3} \times 3)^2R = 3R \quad \text{versus} \quad 1(\text{phase}) \times 3^2R = 9R$$

Moreover, because the single-phase motor operates at a generally lower power factor than the three-phase motor, the actual input for the same horsepower output may be larger than that for the three-phase motor.

Despite these greater losses, the efficiency of a single-phase motor at rated speed and load may be as high as 85%, which compares with about 90% for a three-phase induction motor of the same rating. Nevertheless, because of the generally simpler construction and the need for fewer supply line conductors (and associated equipment, breakers, starters, etc.), the single-phase motor is economically competitive with the three-phase motor. It stands almost alone in the manufacture of very small fractional-horsepower motors.

CONTROL: STARTING AND SPEED CHANGES

The alternating-current series motor has no torque at starting and requires means for starting. For some large motors, starting may be accomplished using another motor for the purpose. For the greater range of ratings of this type of motor, means for starting are provided within themselves and have been described earlier. Some may be started by connecting them directly across the line within permissible limits of voltage variations of the system to which they are connected. Larger-size motors may be started at a reduced voltage through the use of resistances or autotransformers, similar to the methods used for polyphase induction motors.

Speed control may be accomplished through the introduction of a variable resistance in its supply circuit, or through the use of an autotransformer with suitable taps, as has been described for use with polyphase motors. They may be of both the manual and automatic types. The speed of repulsion start motors may be varied by changing the positions of the brushes on the armature; the speed of the motor as well as the direction of rotation depend on the position of the brush axis.

APPLICATIONS

Except for large motors that are used in buses, trolleys, trains, and locomotives supplied electric power from single-phase overhead catenary lines or third rails, most applications of alternating-current single-phase induction motors are in the small-size category, ranging from very small fractional horsepower to about twenty horsepower as the practical limit. In general, the larger sizes are the series type and include large fans, power tools, and traction equipment.

Split-phase motors are used in a variety of equipment and appliances, such as oil burners, ventilating fans, washers, and similar applications.

Capacitor motors range from fractional horsepower to about ten horsepower and are used to drive grinders, drill presses, refrigerator compressors, and other loads that require high starting torque.

Shaded-pole motors have low starting torque, low efficiency, and high noise levels; their use generally encompasses small fans, clock driving motors, and other light loads.

Repulsion-start motors range from fractional-horsepower size to about twenty horsepower and compete with the capacitor motor in applications. It has the disadvantage of commutator and brushes, but this can be used to advantage for speed control as well as change in direction of rotation.

The alternating-current series motor finds application in smaller-size so-called universal motors that can also operate on direct current. They are used in fans, portable tools, and bench tools; in larger sizes, they find application in the transportation field, providing power for traction equipment where electric supply is limited to single-phase. Their employment of commutator and brushes make them less desirable for some applications.

REVIEW

- A single-phase induction motor is somewhat similar to a polyphase squirrel-cage motor except that it has only one stator winding.
- Having no rotating magnetic field at the start, it will have no starting torque. There are several ways of producing an out-of-phase magnetic field that will interact with the stator magnetic field, producing sufficient torque to cause the motor to start turning and continue to turn.
- The split-phase motor has a stator composed of two windings, the axes of which are displaced 90 electrical degrees. When energized, the two windings will produce a rotating field that will rotate around the stator at synchronous speed. Current induced in the squirrel-cage rotor produces a magnetic field that interacts with the rotating magnetic field producing a torque causing the rotor to turn. As the speed of the rotor approaches synchronous speed, the starting winding is automatically disconnected from the line supply (Figure 5–1).
- The capacitor motor is an enhanced version of the split-phase motor. The capacitor in series with the starting winding produces a great phase displacement of the currents in the two windings, resulting in a greater starting torque. When the motor reaches rated speed, the starting winding and capacitor may be disconnected, but may be left in the circuit (Figure 5–2).
- The shaded-pole motor employs a squirrel-cage rotor with modified pole pieces that accommodate a short-circuited winding around a portion of the pole face. As the magnetic field of the pole is increasing, the shading coil is cut by it producing a second magnetic field that acts on the first winding tending to oppose its changes in its alternating cycle. The result is a progressive shift of the magnetic field from the unshaded part of the pole to the shaded part. The conductors of the squirrel-cage rotor are cut by this shifting field, producing a magnetic field that interacts with the

shifting field, causing the rotor to turn. It is the only type of single-phase motor that cannot change its direction of rotation (Figure 5–3).

- The repulsion-start motor has a wound-type rotor with commutator and brushes. Rotor currents are induced by transformer action and the interaction between the stator magnetic field and that produced in the rotor is obtained by shifting the brushes so that the axis of the rotor magnetic field will be at right angles to that of the stator. As the motor approaches synchronous speed, the brushes are removed from the commutator and replaced with a short-circuiting ring around the commutator. The motor then acts as a squirrel-cage induction motor (Figure 5–4).

- The series motor will operate on either direct or alternating current. When operating on alternating current, the alternating magnetic fields of both rotor and stator will interact to produce rotation of the motor. The inductive effect of the alternating magnetic fields on both the stator and rotor will cause distortion of the fields, resulting in low power factor, severe iron losses, and sparking at the commutator. These can be mitigated by embedding a compensating winding in the pole pieces connected in series with the armature. The magnetic field produced by the compensating winding acts to prevent distortion of the main magnetic fields. A similar effect takes place if the compensating winding is short-circuited on itself (Figure 5–5).

- Fractional-horsepower series motors, capable of operating on either direct or alternating current, are called universal motors.

- Losses of a single-phase induction motor are larger than those of a comparable polyphase induction motor. Despite the greater losses, the efficiency of the single-phase motor is only slightly less than that of the polyphase unit. The simpler construction and the need for only a single-phase supply makes it competitive with the polyphase unit. It stands almost alone in the manufacture of fractional-horsepower motors.

STUDY QUESTIONS

1. In the split-phase single-phase motor, how is a rotating field created in the stator?
2. After the split-phase motor comes up to speed, how is the rotating field maintained?
3. Why is the starting torque of a capacitor motor greater than that of the split-phase motor?
4. How may the direction of the split-phase and capacitor motors be reversed?
5. In the shaded-pole motor, what causes the magnetic field to sweep across the pole face from the unshaded to the shaded portion of the pole?

6. What are some disadvantages of the shaded-pole motor?
7. How does the repulsion-start induction motor operate in the starting of the motor?
8. What are the disadvantages of a series motor when operating on alternating current compared to its operation on direct current?
9. What is a compensating winding in the series motor, and what is its function?
10. What are some of the advantages of single-phase motors in the small sizes compared to polyphase motors?

chapter 6

Construction and Maintenance

INTRODUCTION

Motors convert electrical energy into mechanical energy or work. They are the principal means of utilizing mechanical power in factories and offices, on the farm, and in the home. They vary in size from thousand horsepower giants to very small fractional horsepower: from driving trains and ships to operating kitchen appliances and clocks. They must be built to accommodate a wide assortment of conditions in various environments.

BASIC MOTOR TYPES

Motors may operate on direct and alternating current. The conversion to mechanical work is accomplished through the interaction of magnetic fields produced by the field winding and in the winding of the armature; either of these may constitute the stationary part or stator of the motor, and the other the rotating part or rotor. In many instances, the armature windings are contained in the stator, the advantage being that it can be connected to its electric circuit through fixed terminals or contacts rather than through movable or sliding contacts. As these windings may operate comparatively high voltages and carry relatively

Figure 6–1 Types of motors classified according to enclosure: (a) open-type; (b) drip-proof; (c) enclosed. (Courtesy U.S. Navy.)

large currents, connection arrangements are simplified and construction less expensive. Field windings are usually subjected to lower voltages and smaller currents, and slip rings and brushes are usually adequate.

The mechanical power is provided by the rotating shaft of the rotor and utilization devices connected to it directly, through a system of gears or pulleys, or through magnetic or flexible couplings. In any case, the rotor must be designed to sustain the stresses imposed on it during starting and running conditions, as well as during periods of overload. The stator must also be able to sustain the stresses imposed on it reacting to the forces developed that turn the rotor.

Motors may be of the open type for general use, of the semienclosed type for protection against dripping and similar hazards, and of the totally enclosed type when air and water tightness is a requirement (Figure 6–1).

PRINCIPAL MOTOR PARTS

Motors consist of assemblies that generally include some or all of the following elements:

1. A steel frame, yoke, or case that contains pole pieces, field or armature windings, and supports for the rotor shaft bearings
2. Pole pieces and field windings
3. Armature and windings
4. A rotor consisting of a group of copper (or aluminum) conductors mounted in a slotted steel cylindrical core
5. Slip (or collector) rings that provide electrical connection between conductors in the rotating rotor and external electrical circuitry
6. A commutator for changing the direction of the current supplied to or induced in the conductors of the rotor
7. Brushes and holders that provide a contact connection between slip rings or commutator segments to circuits associated with the rotor
8. Bearings for the rotor shaft

Frame, Yoke, or Case

The frame, yoke, or case constitute part of the stator and provides mechanical support for the pole pieces and field windings, or alternatively, for the conductors constituting the armature. End bells or struts are bolted or welded to the frame, yoke, or case and support the rotor shaft bearings. One end bell or strut supports the rigging for the brushes and holders (Figure 6–2), and may also be observed on the motors shown in Figure 6–1. The steel of the frame, yoke, or case provides an excellent path for the magnetic lines of force moving through it, reducing the amperetum requirements of the field.

The arrangement of attachments (armature or field coils, etc.) must be such as to provide a ventilating path for air to circulate to cool the motor. Where a semienclosed case is involved, holes in the casing may be required as part of the ventilating system (Figure 6–3). The case may also contain certain gearing for speed reduction of the motor (Figure 6–4).

Pole Pieces and Field Windings

Stator. For smaller-size motors, the pole pieces, around which are the field windings, are mounted on the inside of the frame, yoke, or case by means of bolts; in larger units, the pole pieces are fitted into grooves, somewhat like a mortise-and-tenon joint. The pole pieces are made of laminated steel, with the laminations insulated from each other by an insulating varnish, to reduce losses from eddy currents (Figure 6–5). They are shaped to fit the curvature of the rotor and the faces modified for shaded-pole motors.

Field coils for smaller units may consist of relatively thin wire wound around the pole pieces. For larger units, the field coil may be preformed and installed on the pole piece and held securely in place between the frame, yoke, or case and the flanged end of the pole. For very small fractional-horsepower

Figure 6–2 (a) End bells bolted to motor frame: supporting rotor shaft and bearings; left also supports rigging for brushes and holders; (b) fabricated bedplate and pedestal support for rotor shaft and bearings. (Courtesy Westinghouse Electric Co.)

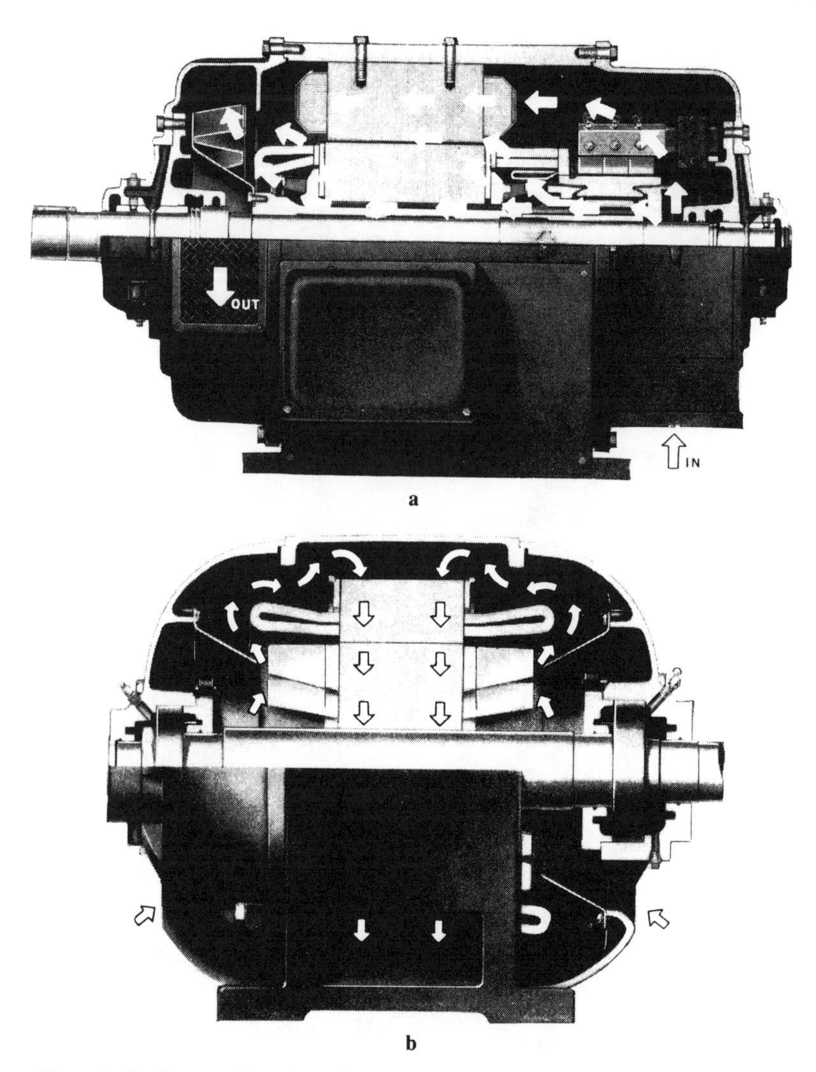

Figure 6–3 Cutaway view of (a) wound-rotor motor and (b) squirrel-cage motor showing methods of ventilation. (Courtesy General Electric Co.)

Figure 6–4 Cutaway view of motor containing gears for speed reduction. (Courtesy General Electric Co.)

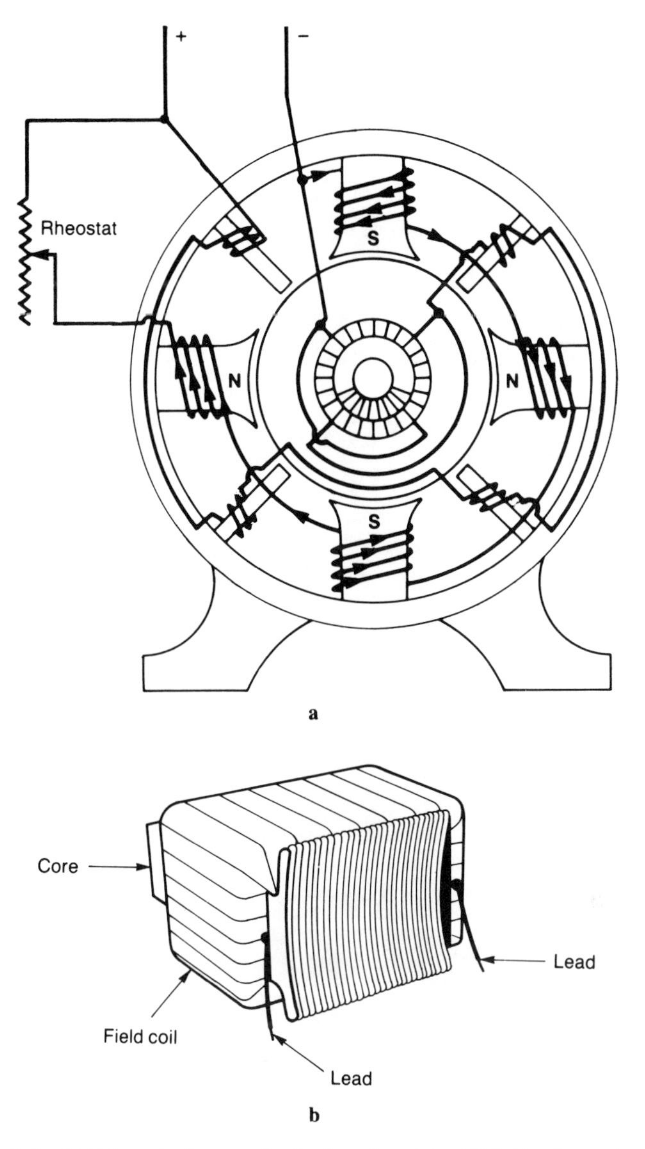

Figure 6–5 (a) Schematic diagram of pole pieces in a stator; (b) field coil on pole piece. (Courtesy Westinghouse Electric Co.)

devices, the case may act as part of the stator and be designed in the shape of and as part of the device.

Rotor. In the case of squirrel-cage rotor fields for some types of alternating-current motors, conductors are embedded in longitudinal slots on the

Figure 6–6 (a) Pole pieces mounted on a rotor for large-size motors; note slip rings for which windings are connected; (b) pole piece showing laminations and dovetail method of mounting; (c) laminated rotor core: note dovetail mounting slots. (Courtesy Westinghouse Electric Co.)

surface of the rotor and are connected together at each end through circular rings or plates. In some instances, one end of the conductors may be connected to slip rings which supply current to the field windings until the rotor comes up to speed, at which point the supply is disconnected and the slip rings connected together to complete the squirrel-cage connection. The rotor is also composed of steel laminations, insulated from each other to restrict eddy current losses.

For large motors, the fixed pole pieces and windings may be mounted on the rotor, fitted into slots provided for that purpose. Leads from the windings are brought out to slip rings. Power is supplied to the field winding by means of brushes that slide on the slip rings (Figure 6–6).

In the case of large synchronous motors, the damper windings (constituting a squirrel-cage rotor) for starting as an induction motor are installed along the periphery of the rotating field pieces, as shown in Figure 6–7.

Windings. The form of the coils and the method of winding and connecting them are influenced by the mechanical shape of the slots: whether they are entirely open or partly closed by the shape of the tooth tips (Figure 6–8). The coils are placed in the slots, with suitable insulation (e.g., varnished paper) to protect them against mechanical injury from the edges and corners of the steel laminations. The coils overlap one another so that each coil has one of its sides in the bottom of a slot and the other on the top of another slot one pitch distant, as shown in Figure 2–4. When the bottom coil is in place, an insulating strip is placed above it and the upper coil is then placed in the slot. A top wedge is driven in so as to hold the conductors in place. The coils are then connected as described in Chapter 2: lap or wave.

Insulation. Coil windings are usually insulated with varnish-impregnated paper or cotton cloth. For high temperatures, insulating materials such

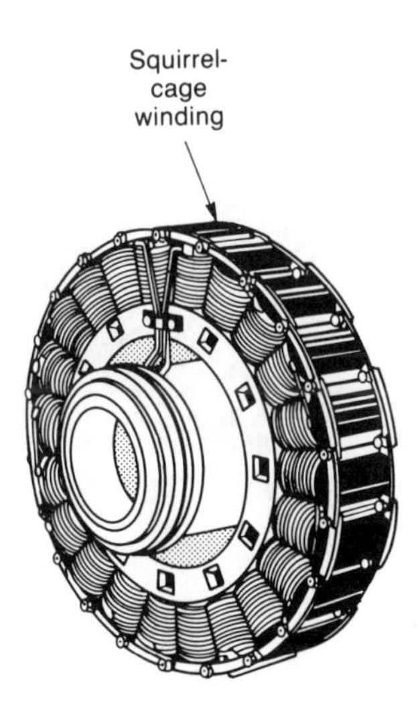

Figure 6–7 Synchronous motor field rotor with starting squirrel-cage or damper windings. (Courtesy Westinghouse Electric Co.)

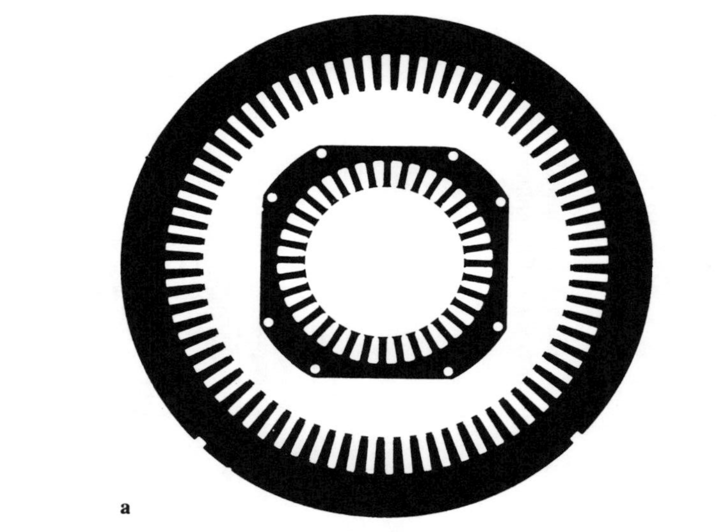

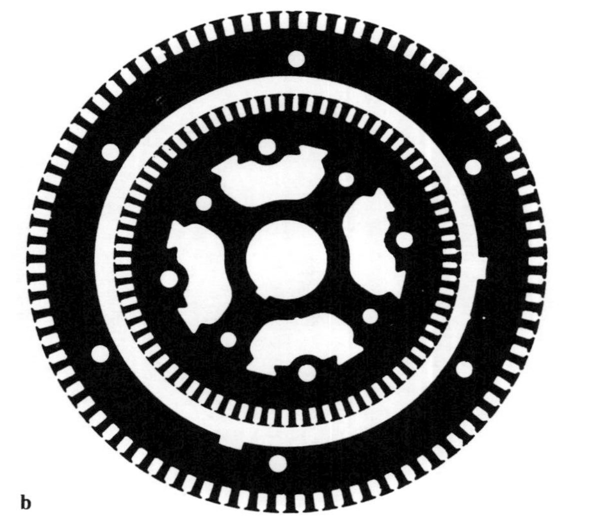

Figure 6–8 Lamination stampings for (a) stators and (b) rotors. (c) details of stamping teeth. (Courtesy Westinghouse Electric Co.)

as mica and (formerly) asbestos with binding substances are used. The insulation provided is designed not only to withstand the normal voltages to which they are exposed, but include a margin to take care of surges that might occur in the supply. The insulation of the leads from such coils is usually deliberately made weaker so that, should failure occur, it may be more apt to take place outside the stator or rotor where access for repairs is more readily available. Insulation for the commutator is usually of mica or paper, but may be a combination of the two.

Armature

Stator. For large motors, the armature conductors are mounted inside the frame, yoke, or case. Preshaped coils are fitted and connected according to the type of motor. The free ends are brought out to terminals. Care is employed in installing the conductors to ensure a smooth and continuous output of the motor (Figure 6–9).

Rotor. The rotor core is made of steel laminations, insulated from each other, as described earlier. In small motors, they are keyed directly to the shaft. In larger machines, they are assembled on a spider and keyed to the shaft. Radial ventilating spaces are provided by inserting spacers at intervals between the laminations and by longitudinal holes in the core. The outer surface is slotted, into which armature coils are inserted and secured. The armature coils are form-wound to the correct size and shape. Sheets of insulating material are placed in the slots to provide additional insulation between the core and the windings. Fiber wedges are driven into the tops of the slots to secure the windings. The free ends of the armature coils are connected to the commutator segments, or to the slip rings, mounted on the same shaft (Figure 6–10). For smaller motors, the armature may consist of conductors placed longitudinally on a rotating steel core, described earlier. A fan is sometimes provided on one end of the shaft to aid in circulating air for ventilation of the motor.

Slip Rings

These are rings made of brass or bronze, but sometimes of steel, mounted on one end of the rotor shaft, but insulated from the shaft. There may be two or three such rings, and their dimensions depend on the current they are designed to carry, which, in turn, depends on the type of motor and whether the rotor constitutes a field or an armature. The ends of the rotor coils are brought out and connected to the slip rings. Slip rings are also referred to as collector rings (Figure 6–11).

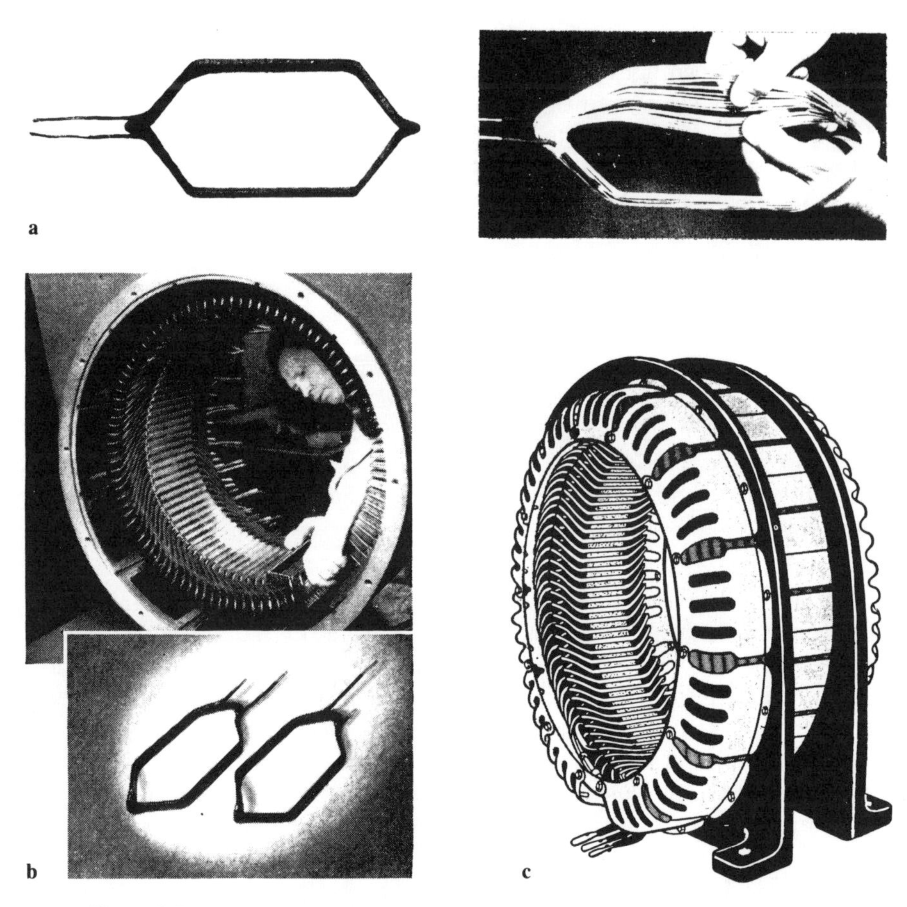

Figure 6–9 Installation of coils in stator: (a) preformed coils; (b) installing coils in stator frame; (c) stator with coils installed. [(a) Courtesy Westinghouse Electric Co.; (b) courtesy General Electric Co.; (c) courtesy U.S. Navy.]

Commutator

A commutator is made up of a number of wedge-shaped bars or segments, usually of hard-drawn copper, assembled into a cylinder and held together by flanges or collars (Figure 6–12a). The segments are insulated from each other by sheet insulation of mica or plastic (Figure 6–12b). The entire commutator is insulated from the supporting rings on the shaft. Connections are made from

Figure 6–10 Types of rotors: (a) wound rotor with commutator; (b) squirrel-cage bar rotor; (c) die-cast squirrel-cage rotor (bars and end rings cast in one piece and in place); (d) wound rotor with collector or slip rings; (e) double-winding squirrel-cage rotor. [(a) Courtesy U.S. Navy; (b)–(e) courtesy Westinghouse Electric Co.]

a
b
c
d
e

Figure 6–11 Slip or collector rings for induction motor. (Courtesy Westinghouse Electric Co.)

the coils on the rotor to the several segments of the commutator (Figure 6-12c and d). To prevent flashover between adjacent segments, commutator and coil designs limit the voltage between the adjacent segments to about 20 V or less. Undercutting the insulation between segments by a small amount (about $\frac{1}{64}$ to $\frac{1}{8}$ in., depending on the diameter of the commutator) allows for pressure of the brushes and obtains better contact, with less tendency to sparking, and less noise (Figure 6–12e). Like the slip rings, the commutator also provides a means of making connection between rotating terminals and stationary external ones.

Brushes and Brush Holders

Brushes carry current from slip rings or commutators to stationary external circuits. They are made of substances that have a relatively low resistance and slide over the moving surfaces of the slip rings or commutator with as little friction as practical. They may be made of a mixture of carbon and graphite, or mixtures of graphite and metallic particles (which also provide a measure of lubrication); they may also be made of rolled copper screening for better conductivity. Within limits, the brushes may be free to slide within their holders to provide for small irregularities that may exist in the curvature of the slip rings or commutator; excessive play, however, may cause sparking, vibration, and misalignment of the brushes with the axis of the commutator or slip rings (Figure 6–13).

To obtain a good contact, pressure of the brushes against the slip rings

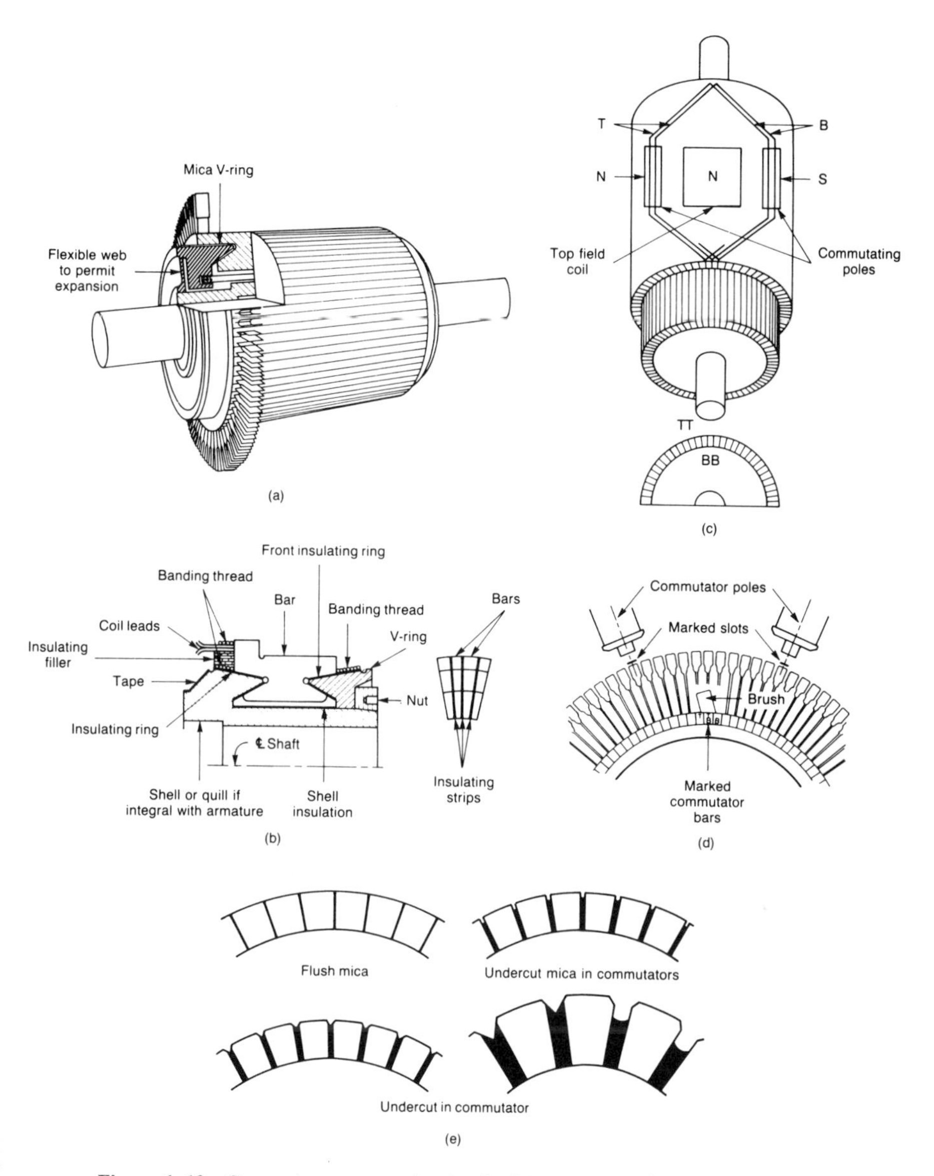

Figure 6–12 Commutator construction details: (a) cutaway section; (b) section details; (c) and (d) commutator and armature markings for locating mechanical neutral; (e) undercuts. [(a), (b) courtesy Westinghouse Electric Co.; (c)–(e) courtesy General Electric Co.]

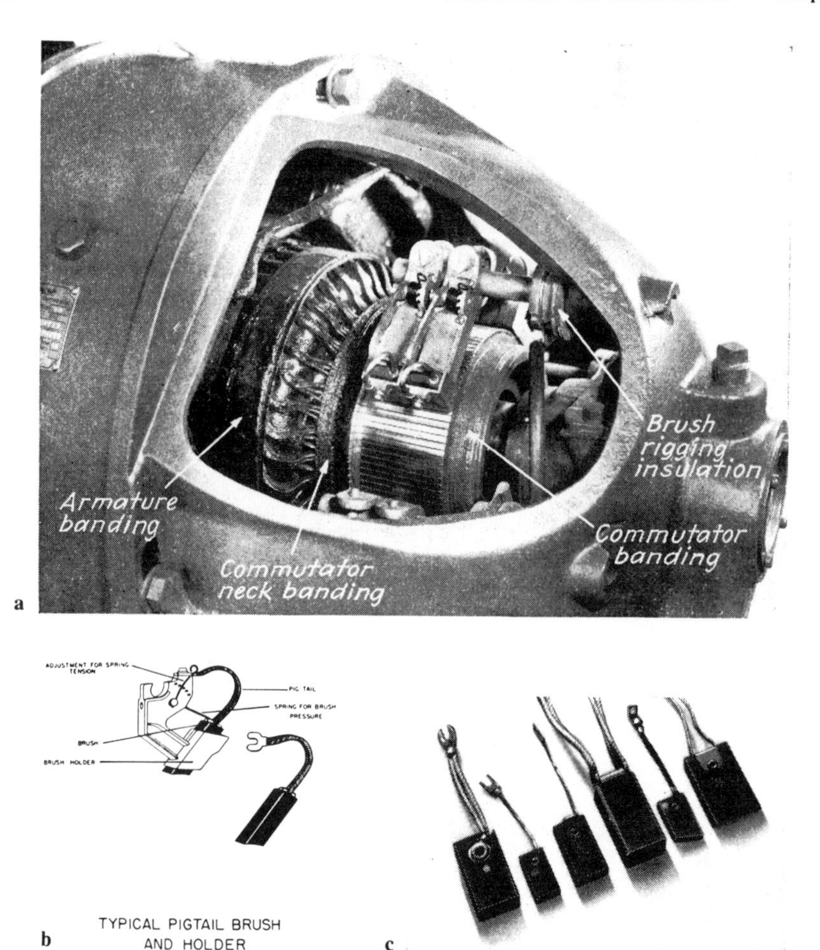

Figure 6–13 Typical brush and brush holder construction and installation: (a) brush details (and commutator) of direct-current motor; (b) typical brush holder; (c) typical pigtail brushes. [(a) Courtesy Westinghouse Electric Co.; (b) courtesy U.S. Navy; (c) courtesy General Electric Co.]

or commutator is usually maintained by springs; pressure should usually not exceed about three pounds per square inch for commutators and about five pounds for slip rings for good contact and reasonable wear of the brushes. Braided copper wire (pigtails) attach each brush to the brush holder and provide a low-resistance connection between them. Different kinds of brushes meet the different conditions in which motors are installed. The brush holders are attached to studs that hold the brushes in their proper positions on the slip rings or commutators. The studs are fastened to a rocker arm that is attached to, but insulated from, the motor frame, yoke, or case. The number of brushes depends on the type of motor in which they are installed. Two or more brushes may be

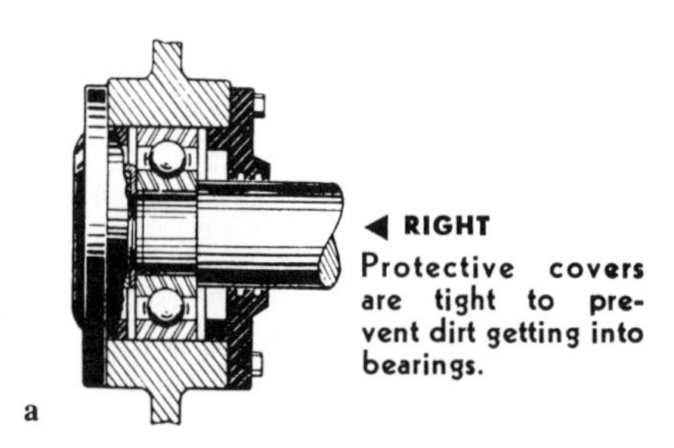

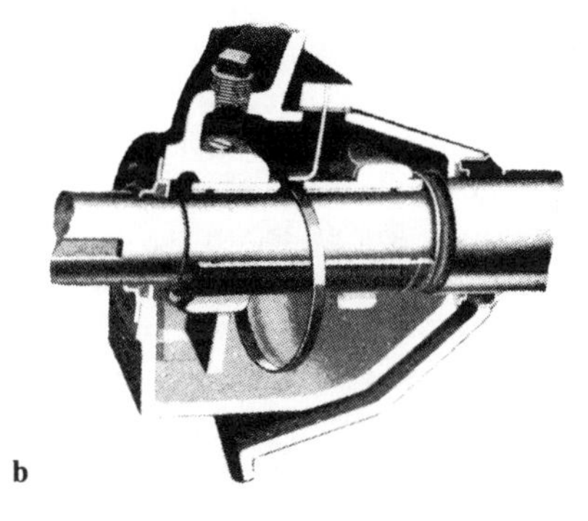

Figure 6–14 Typical motor bearings sealed to keep out dust and dirt: (a) ball bearing; (b) sleeve bearing. (Courtesy General Electric Co.)

connected in parallel for each slip ring or commutator position to provide a greater current-carrying capacity to prevent overloads and overheating of the brushes and holders. They should have sufficient cross-sectional area and be of sufficient number to carry the maximum current of the motor.

Bearings

Depending on the size of motor and its application, bearings to accommodate the shaft of the rotor may be of the sleeve, ball bearing, or roller bearing types. The selection of these and their maintenance have serious influence on the maintenance of the air gap between the stator and the rotor; bearing condition may be determined by checking changes in the air gap (Figure 6–14).

MAINTENANCE

Motors ordinarily found in factories, offices, and homes do not demand extraordinary care to keep them operating efficiently. They should, however, receive planned attention, generally contained in some kind of preventive maintenance program.

Such programs include regular inspections, examinations and tests, adjustments, and servicing. How often these should be done depend on the varying conditions under which the motors operate as well as the type of motors involved and the importance of continued operation. Cursory inspections may be made daily, while disassembling the motor for thorough cleaning, testing, and "group" replacement of parts may be done perhaps every five years as a maximum or more often based on experience.

Daily inspections may include observations of changes in noise levels, odors, vibrations, discoloration, dripping of oil or water, grease spots, and of material piled alongside the motors that may interfere with ventilation or proper operation, or constitute fire hazards.

Weekly, monthly, or quarterly checks may include inspection of brushes and commutators for signs of sparking or heating, windings for signs of dirt or moisture, and corrective measures such as dressing or replacing brushes, sanding or cleaning commutator segments, lubricating and changing oil or grease in bearings, general cleaning of windings, checking air gap and rotors for signs of scraping or of loose electrical connections, and checking of motor starting for time to accelerate to proper speed each time power is applied.

Longer-term checks include resistance tests of winding insulation, grounds, worn bearings, loose or broken connections, eccentric air gap, as well as the items checked more frequently: checks for changes in loads imposed on the motors and adequacy of the motor. Motors may be disassembled and reassembled, with parts replaced even if not completely worn (group replacement). Refer to the manufacturer's instructions and recommendations. Keep records of conditions found, work done, and conditions left.

Very small fractional-horsepower devices, relatively inexpensive, may be discarded and replaced with new units or appliances, depending on the economics of the situation.

Where commutators exist, an indication of possible trouble, fault, or incipient fault in the windings of the rotor may be sparking at the brushes. Sparking or burning of the next commutator bar or segment (in the direction of rotation of the rotor) ahead of the brushes usually indicates the presence of a loose or high-resistance connection, possibly an incipient failure of the insulation, between the rotor coil and the commutator (Figure 6–15).

Insulation

Insulation is perhaps the most vulnerable element of a motor. In addition to the indications provided by the senses (mentioned above), the condition of the insulation may be more definitely and accurately determined by tests. This may be accomplished by measuring the resistance of the insulation by two methods (Figure 6–16): use of an ohmmeter (commonly known as a *Megger*); and by use of an ammeter (or preferably a milliammeter, capable of measuring thousandths of an ampere). Resistance is measured between windings or coils and between windings and ground (the metallic parts—frame, yoke, or case of the motor) and applies to both stator and rotor.

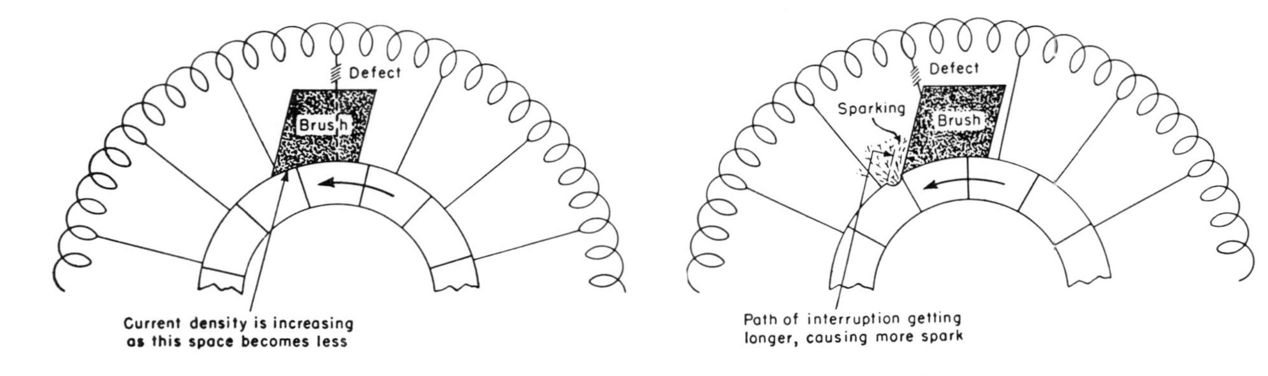

Figure 6–15 Burning commutator symptom of defective connection.

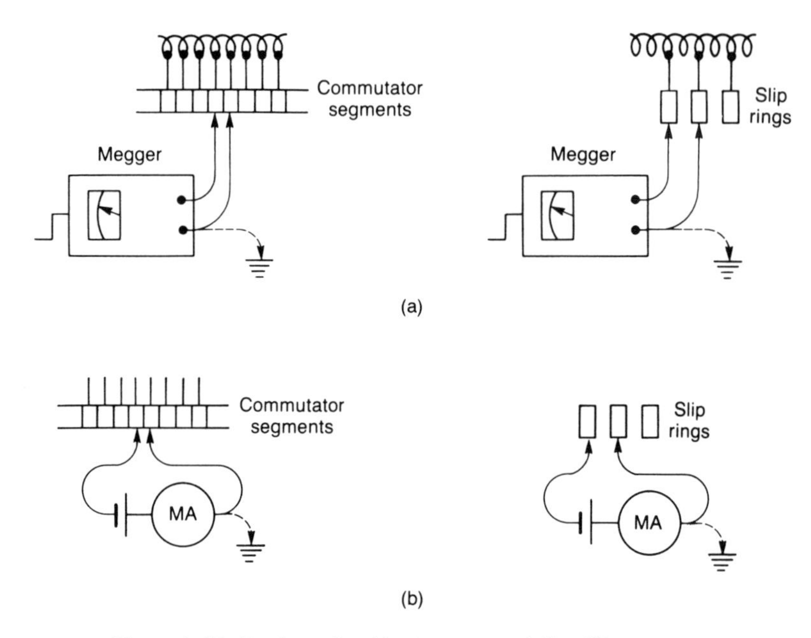

Figure 6–16 Testing coils with (a) megger and (b) milli-ammeter.

For the stator, the measurements are taken between the terminals of the motor, and between them and the frame, yoke, or case of the motor for all motors.

For the rotor, the measurements are taken between the slip rings and between them and the frame, yoke, or case of the motor for alternating-current induction motors. For direct-current and alternating-current synchronous motors, the measurements are taken between the segments of the commutator and between the segments and the frame, yoke, or case of the motor.

Where a commutator is involved, grounds, short circuit between windings or coils, and open circuits can readily be pinpointed.

Where the windings are connected between rotor slip rings or between stator terminals, the location of grounds may be determined by applying a predetermined voltage between two of the slip rings or terminals (at a time). The voltage is then measured between each slip ring or terminal and the frame, yoke, or case of the motor (Figure 6–17). The ratio of the latter voltage (between the slip ring or terminal and the frame, yoke, or case) and the predetermined

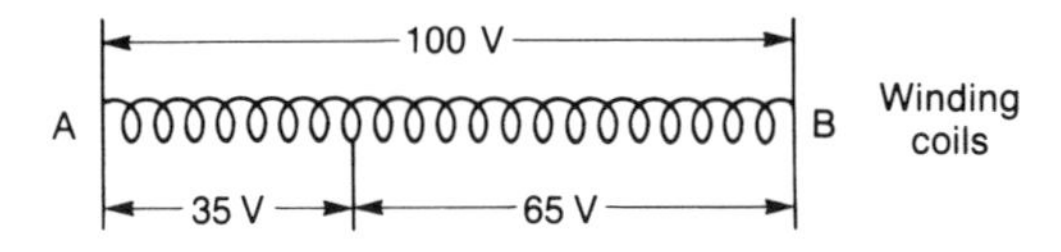

Figure 6–17 Method of testing winding for ground.

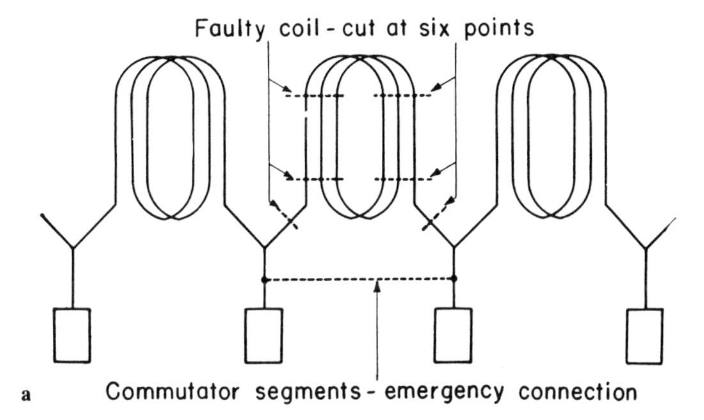

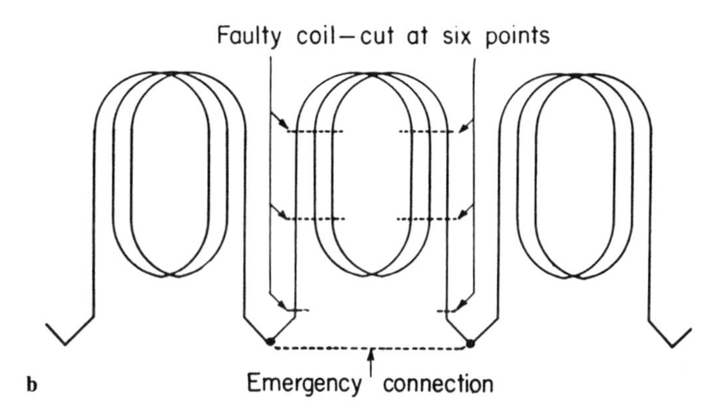

Figure 6–18 Methods of making temporary repairs when winding coil is faulty: how to make an emergency reapir on (a) a motor with a commutator and (b) a motor that has no commutator, so that the motor can be returned to service. These are temporary expedients only. (Courtesy General Electric Co.)

voltage (applied between the slip rings or terminals) will give an approximate but practical indication of the location of the ground. For example, in the figure, if a voltage of 100 volts is applied between the terminals A and B, and the reading between A and ground is 35 volts, that between B and ground 65 volts, the location of the ground is 35/100 of the way from point A and 65/100 of the way from point B.

By comparing the current flow in each of the windings between slip rings or terminals, determination can be made of the existence of short circuits between the coils of the windings. It may be necessary in some instances to disconnect the individual coils of a winding to pinpoint the short circuit.

Should no current be detected during the tests, it can be assumed that one (or more) of the coils may be open circuited. A similar coil-by-coil test may be necessary to pinpoint the open circuit.

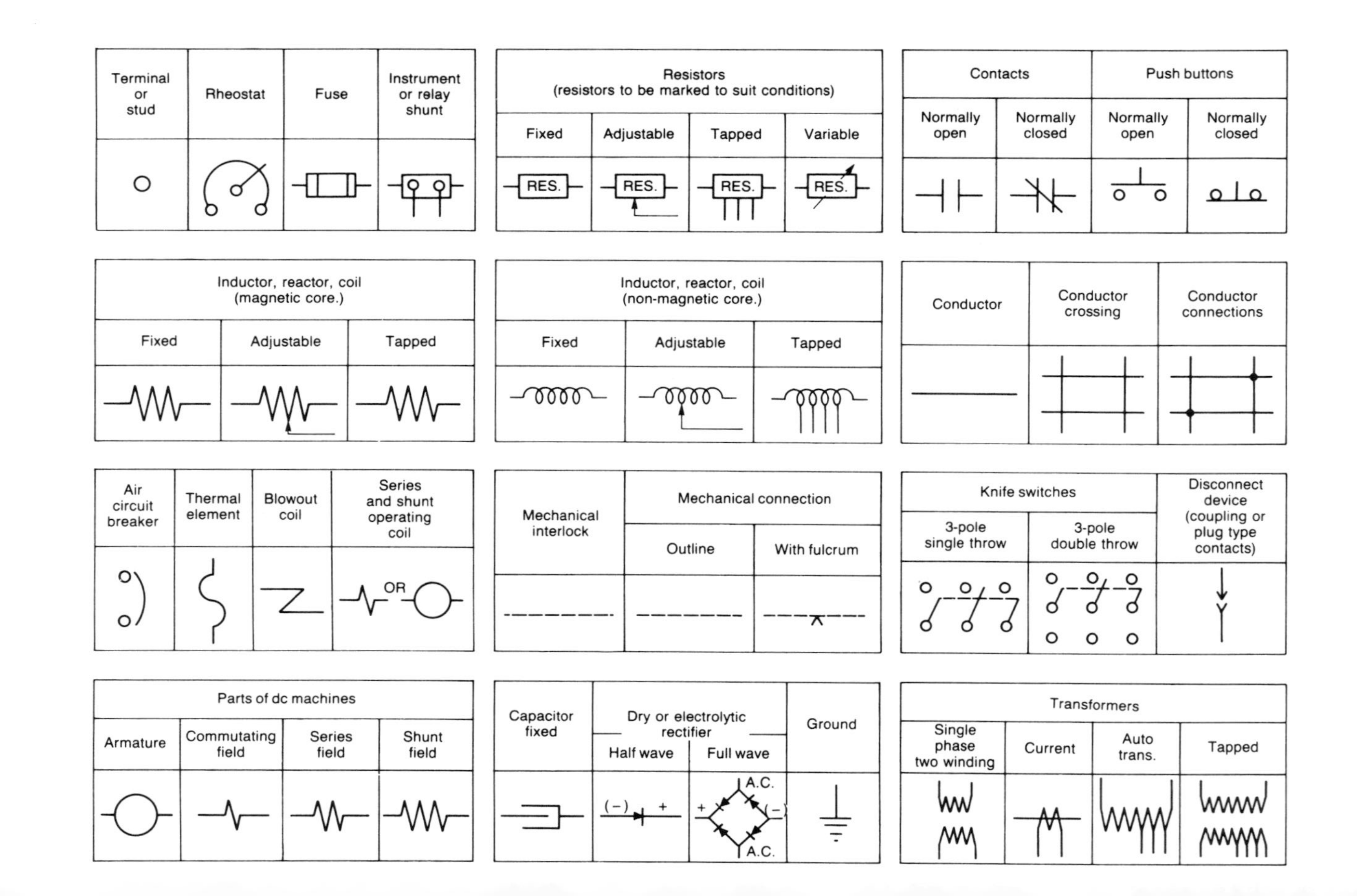
Terminal or stud
Rheostat
Fuse
Instrument or relay shunt
Resistors (resistors to be marked to suit conditions)
Fixed
Adjustable
Tapped
Variable
RES.
RES.
RES.
RES.
Contacts
Push buttons
Normally open
Normally closed
Normally open
Normally closed
Inductor, reactor, coil (magnetic core.)
Fixed
Adjustable
Tapped
Inductor, reactor, coil (non-magnetic core.)
Fixed
Adjustable
Tapped
Conductor
Conductor crossing
Conductor connections
Air circuit breaker
Thermal element
Blowout coil
Series and shunt operating coil
OR
Mechanical interlock
Mechanical connection
Outline
With fulcrum
Knife switches
3-pole single throw
3-pole double throw
Disconnect device (coupling or plug type contacts)
Parts of dc machines
Armature
Commutating field
Series field
Shunt field
Capacitor fixed
Dry or electrolytic rectifier
Half wave
Full wave
(–)
+
A.C.
+
(–)
A.C.
Ground
Transformers
Single phase two winding
Current
Auto trans.
Tapped

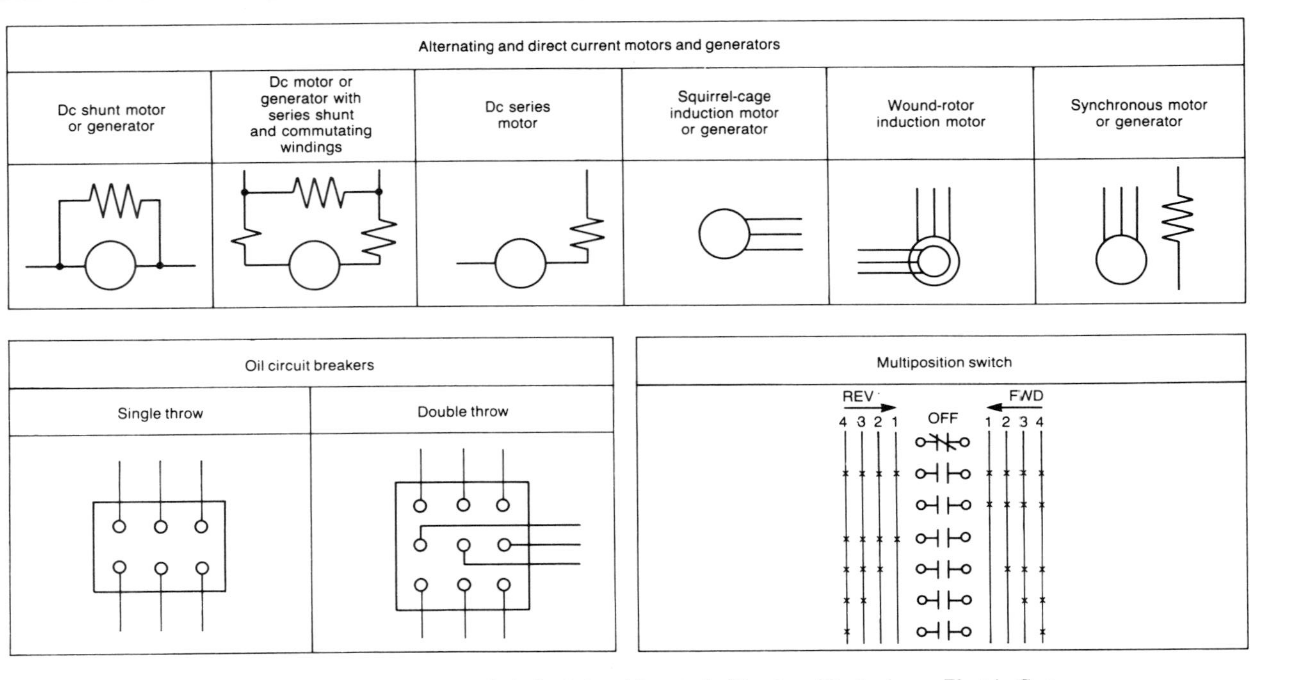

Figure 6–19 Fundamental symbols for industrial control. (Courtesy Westinghouse Electric Co.)

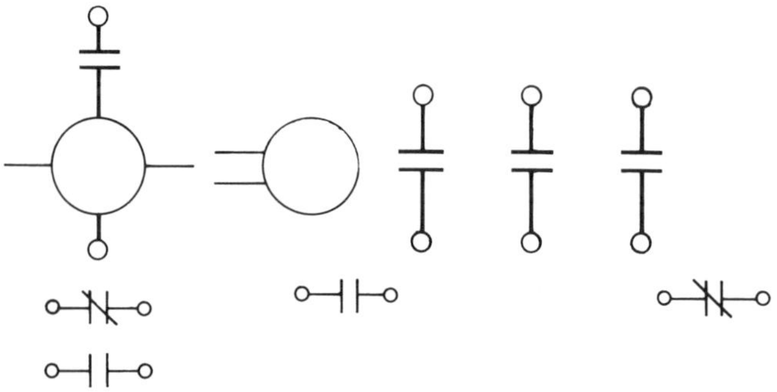

Figure 6–20 Symbols for single-pole and three-pole contactors each with operating coil and with one normally closed interlock. (Courtesy Westinghouse Electric Co.)

The use of direct current is recommended for such tests. The use of alternating current, while feasible, may disturb the observations because of the effect of associated alternating magnetic field in inducing stray currents in the several metallic elements of the motor, resulting in possible false indications.

Records of such periodic tests, their findings, and remedial measures taken should be reviewed as part of any program. Comparisons with past findings may reveal changes in values that may be indicative of changes in the condition of the insulation, enabling preventive measures to be taken before failure should occur.

Diagrammatic presentations of methods to restore the motor to action temporarily by cutting out defective coils is shown in Figure 6–18. As such measures affect the operation of the motor, permanent repairs should be made as soon as possible. For a guide to diagrammatic symbols, see Figures 6–19 and 6–20.

TROUBLESHOOTING

Troubleshooting tips for the various types of motors are listed in Tables 6–1 to 6–3.

TABLE 6–1 Direct-Current Motors

Trouble	Causes	Remedy
Motor will not start	Open circuit in control	Check control for open starting resistor, open switch, or burned fuse
	Low terminal voltage	Check voltage with nameplate rating
	Frozen bearing	Replace bearing, repair shaft if necessary
	Overload	Reduce load or use larger motor
	Excessive friction	Check bearing lubrication
		Disconnect motor from driven machine, turn motor by hand to ascertain whether trouble is in motor
		Disassemble motor, check part by part
		Repair or replace bent shaft
Motor stops after running short time	Motor not getting power	Check voltage at terminals, fuses, overload relay
Motor attempts to start but overload relays operate	Motor started with weak or no field	Check rheostat for correct setting; check condition of rheostat
	Motor torque insufficient to drive load	Check line voltage; use larger motor or one with suitable characteristics to match load
Motor runs too slow under load	Line voltage too low	Check voltage; remove excess resistance in supply or control
	Brushes ahead of neutral	Set brushes on neutral
	Overload	Check load does not exceed allowable motor load
Motor runs too fast under load	Weak field	Check for resistance in shunt field; check for grounds
Sparking at brushes	Commutator in bad condition	Clean and reset brushes
	Eccentric or rough commutator	Grind and true up commutator; undercut mica
	Excessive vibration	Balance armature; check brushes and holders
	Brush and holder acting sluggish	Replace spring, adjust pressure to specifications
	Brushes too short	Replace brushes
	Machine overloaded	Reduce load; use larger motor
	Short circuit in armature	Check and clean armature; check for short circuits in commutator and armature
Noisy brushes	Excessive clearance of brush holders	Adjust holders
	Incorrect angle of brushes	Adjust to correct angle
	Incorrect brushes	Replace with specification brushes

TABLE 6–1 (*Continued*)

Trouble	Causes	Remedy
	High mica	Undercut mica
	Incorrect brush-spring pressure	Adjust to correct value
Uneven brush loading	Brush spring pressures uneven	Adjust pressures; check brush play in holders
Excessive sparking	Poor brush contact	Dress brushes, polish commutator; clean holders
	Wrong pressure on brushes	Check and adjust brushes, springs, and holders
	Brushes off neutral	Set brushes on neutral
	Poor armature connection to commutator	Check armature coils for grounds, incipient fault
Hot field coils	Short circuit between turns	Repair or replace defective coils
Hot commutator	Brushes off neutral	Adjust brushes
	Excess brush pressure	Check and adjust brush springs
Hot armature	Motor overloaded	Reduce motor load
	Winding shorted	Check commutator; test for internal shorts; repair
Commutator grooving	Brushes improperly placed	Relocate brushes properly
Excess brush wear	Rough commutator	Resurface commutator, undercut mica
	Excessive sparking	Realign brushes with commutating fields

TABLE 6–2 Alternating-Current Induction Motors

Trouble	Causes	Remedy
Motor will not start	Overload relay trip	Try restarting; check for causes listed below
	No power	Check supply and protection, fuses, relays, breakers
	Low voltage	Check supply and wiring
	Loose connection	Check connections, tighten
	Winding coil troubles	Check for open or short circuits or grounds
	Bearings stiff	Check lubrication; replace bearings
	Driven machine locked	Disconnect motor from load; check driven machine
	Overload	Reduce load
Noisy motor	Motor single-phasing	Stop motor; try restarting—will not restart; check for opening in circuits
	Vibration	Check driven machine; rebalance rotor
	Air gap not uniform	Center rotor and change bearings if necessary
Motor overheating	Overload	Check load, reduce if necessary
	Load unbalance	Check for single-phasing; check for open circuits, fuses, controls
	Poor ventilation	Clean air passages and windings
	Motor stalled	Check driven machine; bearings
	Uneven rotation	Check for shorted or grounded stator or rotor windings
Hot bearings	Defective bearing or bent shaft	Check lubrication; replace bearing; straighten or replace shaft
	Insufficient or improper oil or grease	Add lubricant; drain, flush and refill, if necessary; check for proper lubricant
	Motor tilted	Relevel motor, realign shaft
Wound-rotor motor runs too slowly	External resistance cut out	Supply wires too small; check for open circuit; check brush pressures and contacts
	Excessive vibration	Balance rotor; check causes listed above

TABLE 6–3 Alternating-Current Synchronous Motors

Trouble	Causes	Remedy
Motor will not start	Load too great	Remove part of load
	Automatic field relay not working	Check power supply, contacts, and connections
	Friction high	Check bearings for tightness and lubrication
	Voltage too low	Reduce external resistance
	Single-phasing	Check for open circuit; test, locate and repair; check for short circuit on one phase
Motor will not come to speed	Excessive load	Reduce load
	Low voltage	Check and raise voltage
	Field excited	Check for relay contact to be open and field discharge contactor closed through discharge resistance
Motor fails to pull into step	No field excitation	Check circuit connections, contacts, rheostat settings
	Load excessive	Reduce load
Motor pulls out of step	Exciter voltage low	Increase excitation
	Trouble in field circuit	Check for open or short circuits, and polarity of field
	Low line voltage	Increase if possible, raise excitation
Motor "hunts"	Fluctuating load	Increase or decrease motor field current
Stator overheats	Rotor not centered	Realign, adjust bearings
	Open phase	Check connections and correct
	Unbalanced currents	Loose or improper internal connections; tighten, correct
Coils overheat	Short circuit	Cut out coil; repair or replace
Field overheats	Short circuit	Replace or repair
	Excessive current	Reduce excitation
All motor parts overheat	Overload	Reduce load or increase size of motor
	Over- or underexcitation	Adjust excitation to ratings
	No field excitation	Check circuit and exciter
	Reverse field coil	Check polarity; change leads
	Improper voltage	Check and apply proper voltage
	Improper ventilation	Remove any obstruction, clean out dirt
	Excessive ambient temperature	Supply cooler air, check ventilation; install fans if necessary

REVIEW

- Motors consist basically of a stator and a rotor. For smaller motors, the field winding may be contained in the stator and the armature winding in the rotor. For larger units, the arrangement may be reversed.
- Field windings are usually subjected to lower voltages and currents than those of the armature.
- Motors may be of the open, semienclosed, or dripproof type, and totally enclosed or air- and watertight types (Figure 6–1).
- Open-type motors are ventilated by ambient air currents; semienclosed motors may have fan blades attached to one end of the rotor shaft and strategically placed openings in the rotor and casing to ensure proper ventilation; completely enclosed motors are built to run hotter with no ventilation (Figure 6–3).
- Field, pole piece, and rotor are all made of laminated steel, with laminations insulated from each other, to restrict iron losses (Figures 6–5 and 6–6).
- In squirrel-cage rotors, the conductors may be embedded in the rotor surface without insulation (Figure 6–10).
- In wound-rotor motors, the conductors are insulated and the ends connected to slip rings or commutators (Figure 6–10).
- A commutator is made up of a number of wedge-shaped bars or segments assembled in cylindrical form, each segment insulated from adjacent ones by mica or plastic insulation. The insulation between segments is undercut a small amount to reduce wear, sparking, and noise (Figure 6–11).
- Brushes are made of graphite-carbon or metallized carbon materials, or of copper meshing. To maintain good contact, they may be held against the slip rings or commutators by springs. Brush holders are attached to insulated studs in such a manner as to permit them to accommodate small vertical and horizontal movements (Figure 6–13).
- Insulation of windings may be varnish-impregnated paper or cloth, mica and some plastics, or a combination. Insulation of winding leads may be made deliberately weaker so that failure from voltage surges may take place where it can be more readily discovered and handled.
- Bearings may be of the sleeve, ball, or roller bearing types (Figure 6–14).
- Maintenance may consist of daily, short-time, and long-time periodic inspections. Routine maintenance may include lubrication, dressing or replacement of brushes, and a general cleaning of windings and some

cursory testing of winding insulation and electrical connections. Long-term or preventive maintenance may include disassembling the motor, repair or replacement of parts, and reassembling of the motor.

✦ Routine maintenance depends largely on observation and testing. The senses serve to discover heat, sounds, smell, and feelings that often provide the first warnings of incipient failures. Records serve to indicate changes in conditions occurring between observations and tests.

STUDY QUESTIONS

1. Why is the armature winding of a motor sometimes situated in the stator?
2. Which windings are subjected to higher voltages and currents, those of the field or armature?
3. What are the principal parts of a motor?
4. What are several types of motor construction? How do they differ?
5. How are the several types of motors ventilated?
6. How is a commutator constructed?
7. What are some insulation materials? Where are they used?
8. What are some of the materials of which brushes are made?
9. What types of bearings are employed to support the rotor shaft?
10. What types of maintenance are employed in servicing motors? How do they differ?

PART B Electricity and Magnetism

chapter 7

Elements of Electricity

ELECTRON THEORY

Modern science has discovered that electricity constitutes a means of transmitting energy through the motion of electrons. To define electricity, therefore, it is necessary to describe electrons and their behavior.

Electrons

Every substance is said to be made up of *atoms*. Each atom is a miniature solar system consisting of a ''sun'' and ''planets'' revolving about it (Figure 7–1). These ''planets'' are known as *electrons* and are identical in every atom, no matter what the substance may be. The atoms of different substances, however, each have different numbers of electrons normally revolving in their solar systems; what makes them different substances are the compositions of the ''sun'' (called protons) about which the electrons revolve.

The electrons can be removed from the solar systems of their atoms without changing the character of the substance. But just as soon as one is removed, another electron from an adjacent atom takes its place. That atom in turn acquires another electron from the next adjacent atom. The process continues so that the number of electrons in each atom tends to remain constant. This shifting or movement of electrons is what is known as the *conduction of electricity*.

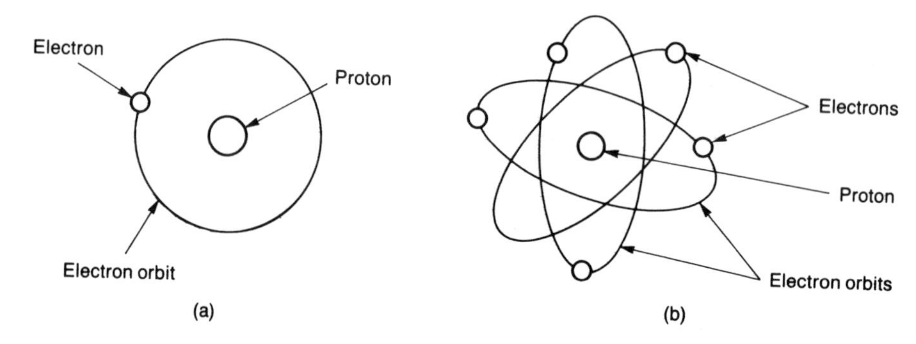

Figure 7–1 (a) Simple atom; (b) atom with five electrons.

Electric Charges

It will be observed that when an atom loses one or more electrons, it has a strong tendency to attract electrons away from adjacent atoms. It sometimes happens that the adjacent atoms themselves have lost some electrons, and they, too, are seeking to attract them from other atoms. A condition is reached where the atoms that have lost electrons are unable to replenish the loss, but the attracting force nevertheless continues to exist. When this happens, the substance is said to be *charged*.

The electrons lost by these atoms must go somewhere. They may be pushed by an outside influence into the solar systems of other atoms, from which they cannot return. The latter atoms then have a surplus of electrons and exhibit a strong tendency to throw off these extra electrons. When this happens, it is again said that the substance is charged.

Now, if these two sets of atoms are brought near each other, the attracting tendency of one matches the tendency of the other to cast off the surplus electrons. The result will be that the surplus electrons of one set of atoms move into the solar systems of the other set of atoms and all the atoms will then be returned to their original state. Such an action is also a flow of electricity, even though it may exist only momentarily.

To distinguish these atoms from each other when charged, the atoms that have lost electrons are said to have a *positive* or plus charge, while those that have a surplus of electrons are said to have a *negative* or minus charge. (An atom or group of atoms carrying an electric charge is known as an *ion*.)

CONDUCTORS AND INSULATORS

If a negatively charged body is discharged by connecting it through a wire to the earth, the surplus number of electrons in that body (which gives it the negative charge) pass into the wire, while an equal number of electrons pass

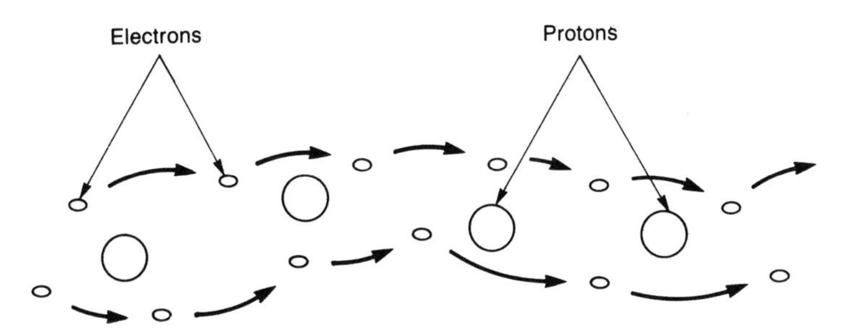

Figure 7–2 Electron flow.

from the wire into the earth. The surplus electrons do not simply race individually through the wire to the earth. The atoms of the wire are already full of electrons. Those coming in, therefore, push along the electrons that are already there until an equal number pass out at the other end to the earth, and the wire is neutral again.

When a positively charged body is discharged to the earth in a similar manner, the process is the same, but the electrons move in an opposite direction. The electron-deficient (positively charged) body draws in electrons from the wire and these are replenished by electrons drawn up from the earth. The same process would result if the two oppositely charged bodies were connected together instead of each being connected to the earth. This process or flow of electrons is called *conduction* (Figure 7–2).

Some substances are better conductors of electricity than others. These are the materials that contain a plentiful supply of electrons and offer little obstruction to the movements of these electrons. Poor conductors are those that contain only a limited number of electrons and do not allow their free movement. All metals are generally good conductors. Nonmetals are usually poor conductors; the poorest of these are termed *insulators*.

ELECTRIC PRESSURE OR VOLTAGE

An electric current generator of any sort is simply a means for setting the electrons in motion. This generator may be likened to a pump which creates an electrical pressure that sends the electrons streaming through the wires. This electrical pressure is expressed in volts (named in honor of the Italian physicist Alessandro Volta). An ordinary dry-cell battery can, through chemical action, create an electrical pressure of approximately $1\frac{1}{2}$ volts.

CURRENT

The flow of electrons constitutes an electric current. The number of electrons that pass a reference point in a second determines the current strength, just as the flow of water is measured in gallons per second. The flow of electricity or current is expressed in amperes (named after the French physicist André Marie Ampère). According to scientific measurement, 6.29 billion billion electrons passing in one second make up one ampere.

RESISTANCE

The wires in an electric circuit can be thought of as an electrical pipe, because the electric current (or stream of electrons) flows through it. It is obvious, then, that the larger the diameter of the wire, and the shorter in length it is, the easier the flow or the less is the electrical resistance. The thinner and longer the wire, the greater is its electrical resistance. The same is true of water pipes, where friction, in place of electrical resistance, impedes the flow of water more in small, long pipes than in large, short ones. To a certain extent, electrical resistance may be thought of as due to friction between the electrons flowing and pushing each other in a conductor. Electrical resistance is expressed in ohms (after the German scientist Georg Simon Ohm). The resistance of about 150 feet of ordinary No. 18 (AWG) copper wire is approximately one ohm.

OHM'S LAW

In any electric circuit, the three factors of pressure, current, and resistance are definitely interrelated. Thus, if the electrical pressure applied to a circuit is increased, more electrons will be set in motion and the current flowing in the circuit will be increased. Similarly, if the resistance to the flow of electrons is increased, that is, by increasing the resistance of the circuit, and the applied voltage maintained at its original value, the flow of electrons—or electric current—would be reduced. Finally, if it is desired to maintain a given flow of electrons—or electric current—in a circuit whose resistance could be varied, then as the resistance is increased, the voltage or electrical pressure would also have to be increased. These three statements may be rewritten in simplified form as follows:

1. Current varies with pressure (or voltage) when resistance is fixed.
2. Current varies in inverse proportion with resistance when pressure is fixed.
3. Pressure varies with resistance when current is fixed.

More simply, the current flowing in a circuit is directly proportional to the electrical pressure and inversely proportional to the electrical resistance. This is known as *Ohm's law*. Thus, if the voltage applied to a circuit and the resistance of the circuit are known, the amperes flowing can be found by dividing the volts by the ohms.

The law may be stated compactly in three forms. Let

I = amperes flowing in the circuit
E = electrical pressure or voltage applied to the circuit
R = resistance of the circuit

Then the law may be written as follows:

$$\text{Form 1: } I = \frac{E}{R}$$

This expression is used when the volts and ohms are known and it is desired to find the current in amperes.

$$\text{Form 2: } E = IR$$

This expression is used when the amperes and ohms are known and it is desired to find the electrical pressure in volts. (The two letters IR written together mean I multiplied by R.)

$$\text{Form 3: } R = \frac{E}{I}$$

This expression is used when the volts and amperes are known and it is desired to find the resistance in ohms.

Thus it is seen that if any two of the quantities above are known, the third can be determined. In practice, the two quantities usually known are the electrical pressure, measured in volts, and the current, measured in amperes. They are measured with voltmeters and ammeters.

DIRECT AND ALTERNATING CURRENT

There is a resemblance between the flow of electricity in an electric circuit and the flow of water in a water system. In Figure 7–3a a simple electric circuit is diagrammatically illustrated. This consists of a generator, motor, switch, and wire conductors. In Figure 7–3b is shown a water system consisting of a pump, waterwheel motor, pipes, and valve, which correspond to this electric circuit.

Consider what happens when the pump is started, driven by some outside source of power. If the valve is turned off, water will be sucked out of pipe A and forced into pipe B until the pump has built up as much difference in pressure between these pipes as it can. If the valve is turned on, the water will at once begin to flow through the pipes, turn the waterwheel motor, and continue to flow around and around the system. The waterwheel motor can then be used to drive machinery. When the valve is turned off, the water will stop flowing and the waterwheel motor will stop, but the difference in water pressure between the two pipes is maintained as long as the pump runs.

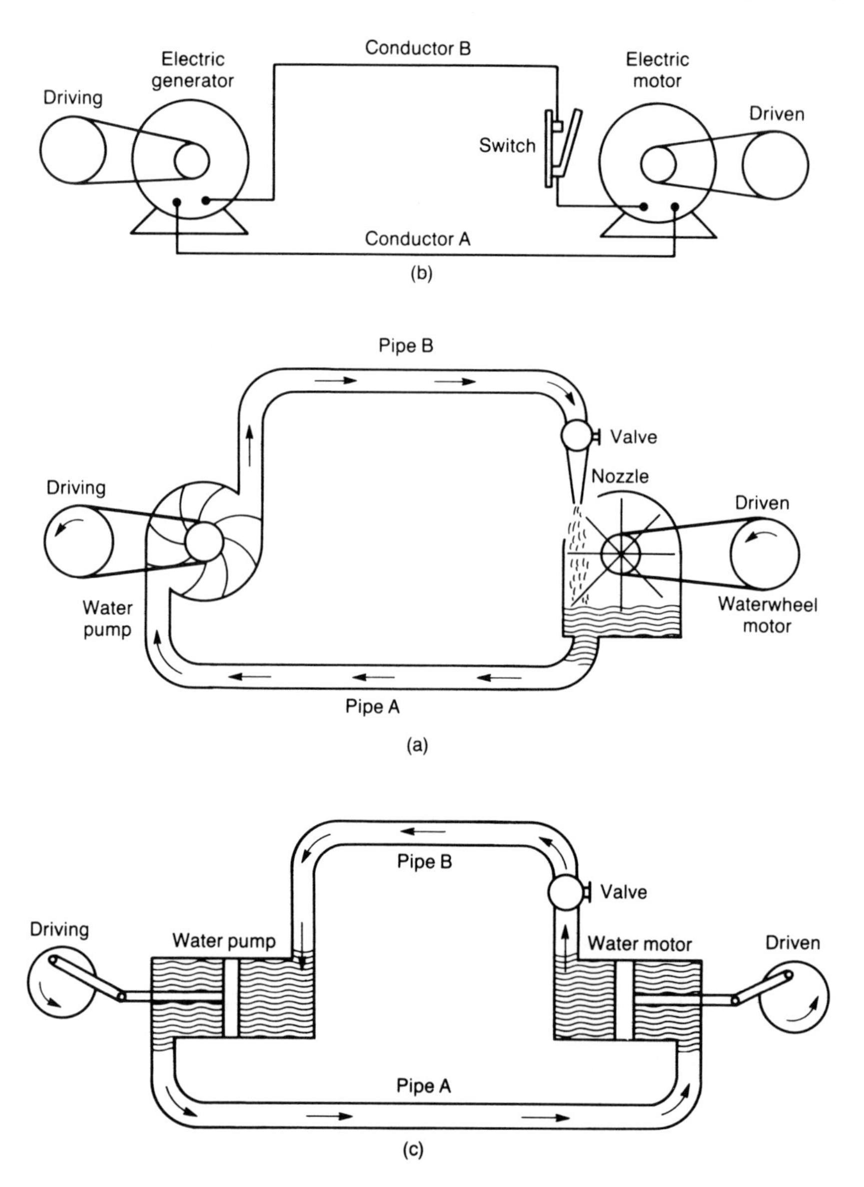

Figure 7–3 Water analogy of direct and alternating current.

Exactly the same thing occurs in the electric circuit. When the generator, driven by an outside source of power, is started, electrons are drawn out of wire A and piled up in wire B against the switch, which is open. This continues until the generator has built up as much pressure—or voltage—between the wires as it can. Now, when the switch is closed or turned on, electrons will flow around the circuit through the motor, which will then operate. When the switch is again opened, the motor will stop because the electrons can no longer flow in the circuit, but the difference in pressure or voltage between the wires is maintained as long as the generator continues running.

Referring now to Figure 7–3c, consider what happens when the piston of the pump is actuated from an outside source of power. Water in pipe B will be pushed in one direction, causing a flow of water or current in that direction. Assuming that the valve is open, the water will push the piston in the water motor in the same direction. Now, as the piston of the pump moves in the opposite direction, water in pipe A will be pushed, causing a flow of water in the direction of the water motor. The water will push the piston of the water motor in the direction opposite to that previously moved. Thus, an oscillating, or to and fro, motion is transmitted from one end of the circuit to the other. The piston water motor can then be used to drive machinery.

It is seen, therefore, that complete circulation of water—or electrons—is not necessary to transmit energy from one point to another. In the latter method, currents alternate their direction periodically, first flowing in one direction and then in the other. In electric circuits, this is called alternating current—ac—to distinguish it from the first method, the continuous or direct current—dc. By an *electric current* is meant a flow of electricity (or electrons), which may be all in one direction (direct current) or back and forth (alternating current), but in any case a continuous performance which can be kept up indefinitely.

REVIEW

- Science theorizes that electricity constitutes a means of transmitting energy through the motion of electrons.
- Every substance is said to be made up of atoms which, in turn, are made up of electrons orbiting about a proton. Electrons can be removed from one atom and transferred to another.
- Atoms that have lost one or more electrons have a strong tendency to attract electrons from adjacent atoms; those that have a surplus of electrons have a strong tendency to throw off the extra electrons. Atoms in these conditions are said to be charged; the atoms that have lost electrons are said to have a positive charge; those with a surplus are said to have a negative charge.
- An electric current is said to constitute a flow of electrons from the negative charged atoms to the positive charged atoms.
- Materials that contain a plentiful supply of electrons and offer little obstruction to the movement of these electrons are good conductors of electricity. Materials that contain only a limited number of electrons and do

not allow their free movement are poor conductors, and the poorest of these are termed insulators.

- ✦ Any means that sets the electrons in motion is termed an electrical pressure and is expressed in volts; this pressure is sometimes referred to as voltage.
- ✦ The number of electrons that flow past a given point in a second is called an electric current and is expressed in amperes; current is sometimes referred to as amperage.
- ✦ Friction between the electrons flowing in a conductor is said to constitute its resistance and is expressed in ohms.
- ✦ The relationship between these three quantities is expressed in Ohm's law, which states the current flowing in a circuit varies directly with the pressure applied and inversely as the resistance of the circuit.
- ✦ Energy can be transmitted by a direct current of electrons flowing from one point to another, or by an alternating current of electrons having a push-pull effect; each method results in a continuous performance of mechanical work.

STUDY QUESTIONS

1. According to theory, what is the structure of the atom?
2. What is an electron?
3. How are the atoms of one substance different from another?
4. What is meant when an atom is said to be positively charged? When negatively charged?
5. What is the relation between the flow of electrons and the flow of electric current?
6. Explain metallic conduction and why some substances are better conductors than others.
7. What are the units of electrical pressure, current, and resistance?
8. On what characteristics does the resistance of a wire depend?
9. Express Ohm's law in the three forms by which current, resistance, and voltage can be determined, respectively. What are direct and inverse proportion?
10. Describe the analogy between the water circuit and the electric circuit, both for direct and alternating currents.

chapter 8

Properties of Electric Circuits

ELECTRIC CIRCUITS

In every electric circuit there must be a complete path for the flow of the current. This path extends from one terminal of the source of supply (usually designated as the "negative" terminal), through one of the conducting wires, through the device or devices using the energy, through the other of the conducting wires to the second terminal of the source (usually designated as the "positive" terminal), and back through the source to the first (or negative) terminal.

There are two fundamental types of electric circuits, known as the series circuit and the multiple or parallel circuit. Other types are a combination of these. The following discussion applies both to devices producing electrical energy (generators) and to devices receiving electrical energy (appliances).

Series Circuits

In a *series circuit*, all the parts that make up the circuit are connected in succession, so that whatever current passes through one of the parts passes through all the parts. The circuit shown in Figure 8–1 contains four resistances, R_1, R_2, R_3, and R_4, which are connected in series. The same current (I) flows through each of these resistances. Assume that an electrical pressure (E) of 100

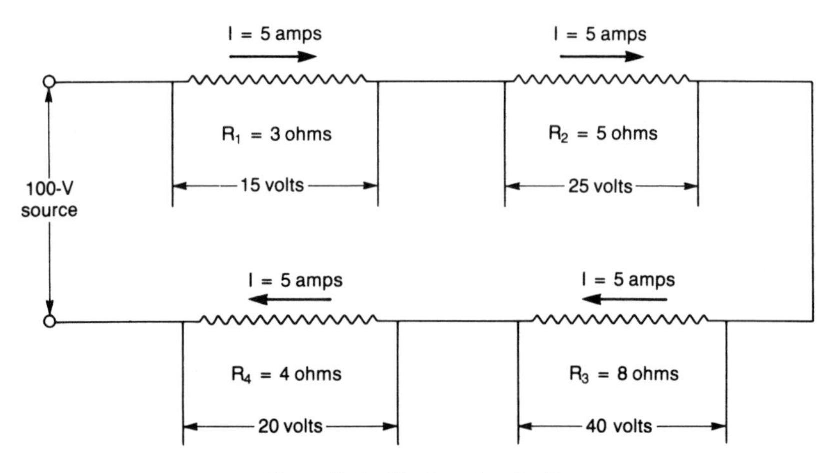

Figure 8–1 Simple series circuit.

V is applied across the terminals of the circuit and that a current of 5 A flows through the circuit. By applying Ohm's law, the resistance (R) of the circuit as a whole is found to be 20 Ω; that is,

$$R = \frac{E}{I} \quad \text{or} \quad \frac{100 \text{ volts}}{5 \text{ amperes}} = 20 \text{ ohms}$$

This total resistance is the sum of the resistances, R_1, R_2, R_3, and R_4. The current of five amperes is the same throughout the circuit. The volts across R_1, R_2, R_3, R_4 are measured and found to be

$$E_1 = 15 \text{ volts} \qquad E_2 = 25 \text{ volts}$$
$$E_3 = 40 \text{ volts} \qquad E_4 = 20 \text{ volts}$$

Now, Ohm's law applies to any part of the circuit as well as to the entire circuit. Apply Ohm's law to each resistance in the series circuit:

$$\text{since } E_1 = 15 \text{ volts;} \quad R_1 = \frac{E_1}{I} = \frac{15 \text{ volts}}{5 \text{ amperes}} = 3 \text{ ohms}$$

$$\text{since } E_2 = 25 \text{ volts;} \quad R_2 = \frac{E_2}{I} = \frac{25 \text{ volts}}{5 \text{ amperes}} = 5 \text{ ohms}$$

$$\text{since } E_3 = 40 \text{ volts;} \quad R_3 = \frac{E_3}{I} = \frac{40 \text{ volts}}{5 \text{ amperes}} = 8 \text{ ohms}$$

$$\text{since } E_4 = 20 \text{ volts;} \quad R_4 = \frac{E_4}{I} = \frac{20 \text{ volts}}{5 \text{ amperes}} = 4 \text{ ohms}$$

As a check, it is found that the sum of the four individual voltages equals the total voltage:

$$E_1 + E_2 + E_3 + E_4 = E$$

$$15 \text{ volts} + 25 \text{ volts} + 40 \text{ volts} + 20 \text{ volts} = 100 \text{ volts}$$

Also, the sum of the separate resistances is equal to the total resistance; that is,

$$R_1 + R_2 + R_3 + R_4 = R$$
$$3 \text{ ohms} + 5 \text{ ohms} + 8 \text{ ohms} + 4 \text{ ohms} = 20 \text{ ohms}$$

When a current flows in a circuit, there is a continual drop in electrical pressure from one end of the circuit to the other. This drop is derived from Ohm's law. It is $E = IR$ and is usually known as the *IR* drop. In all leads and connecting wire, this drop is kept as small as possible because it represents a loss.

In every appliance through which a current is sent there is also an *IR* drop, usually much larger than in the connecting wires because the resistance of the appliance is higher. If several appliances are connected in series, the electrical pressure does not drop evenly around the circuit but by steps, each step representing an appliance. Since the current is the same in all of them, the *IR* drop in each appliance is proportional to its resistance.

The generator or source of electrical pressure provides a certain total voltage for the circuit. Each appliance spends a part of it, but all these parts must add up to the original pressure or voltage.

Multiple or Parallel Circuits

A *multiple* or *parallel circuit* is one in which all the components receive the full line voltage, the current in each part of the circuit being dependent on the amount of opposition (resistance) of that part of the circuit to the flow of electricity.

The circuit shown in Figure 8–2 contains four resistances, R_1, R_2, R_3, and R_4, connected in multiple. Assume again that an electrical pressure (E) of 100 volts is applied across the terminals of the circuit. Here each branch of the

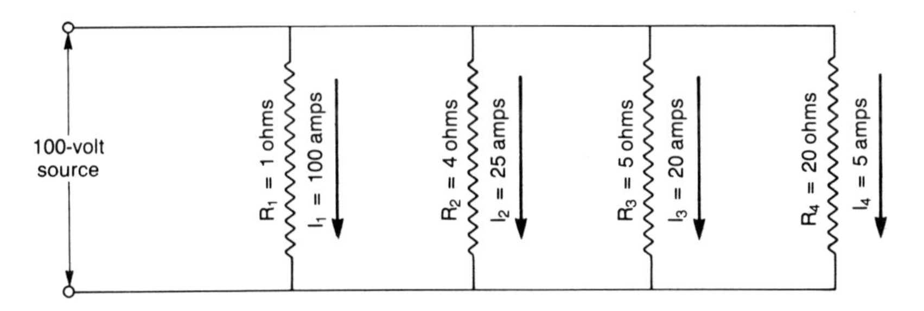

Figure 8–2 Resistances in multiple.

circuit receives the full voltage of 100 volts. Assume that the resistances have the following values:

$$R_1 = 1 \text{ ohm}, R_2 = \cdot 4 \text{ ohms}, R_3 = 5 \text{ ohms}, R_4 = 20 \text{ ohms}$$

Again, apply Ohm's law to find the current flowing in each resistance:

$$\text{since } R_1 = 2 \text{ ohms;} \quad I_1 = \frac{E}{R_1} = \frac{100 \text{ volts}}{1 \text{ ohm}} = 100 \text{ amperes}$$

$$\text{since } R_2 = 4 \text{ ohms;} \quad I_2 = \frac{E}{R_2} = \frac{100 \text{ volts}}{4 \text{ ohms}} = 25 \text{ amperes}$$

$$\text{since } R_3 = 5 \text{ ohms;} \quad I_3 = \frac{E}{R_3} = \frac{100 \text{ volts}}{5 \text{ ohms}} = 20 \text{ amperes}$$

$$\text{since } R_4 = 20 \text{ ohms;} \quad I_4 = \frac{E}{R_4} = \frac{100 \text{ volts}}{20 \text{ ohms}} = 5 \text{ amperes}$$

The total current in a parallel circuit is equal to the sum of the separate currents:

$$I_1 + I_2 + I_3 + I_4 = I$$

$$100 \text{ amps} + 25 \text{ amps} + 20 \text{ amps} + 5 \text{ amps} = 150 \text{ amps}$$

The resistance of the entire circuit may be found by applying Ohm's law:

$$R = \frac{E}{I} = \frac{100 \text{ volts}}{150 \text{ amps}} = \frac{2}{3} \quad \text{or} \quad 0.667 \text{ ohm}$$

Another way of obtaining the resistance of the entire circuit is to add the reciprocals of each of the resistances and taking the reciprocal of the sum (the reciprocal of a number is equal to 1 divided by that number); that is,

$$\frac{1}{R_1} + \frac{1}{R_2} + \frac{1}{R_3} + \frac{1}{R_4} = \frac{1}{R}$$

$$\frac{1}{1 \text{ ohm}} + \frac{1}{4 \text{ ohms}} + \frac{1}{5 \text{ ohms}} + \frac{1}{20 \text{ ohms}} = \frac{1}{R}$$

$$1.00 + 0.25 + 0.2 + 0.05 = 1.50 = \frac{1}{R}$$

The reciprocal of $1/R$ is R or

$$R = \frac{1}{1.50} = 0.667 \text{ ohm}$$

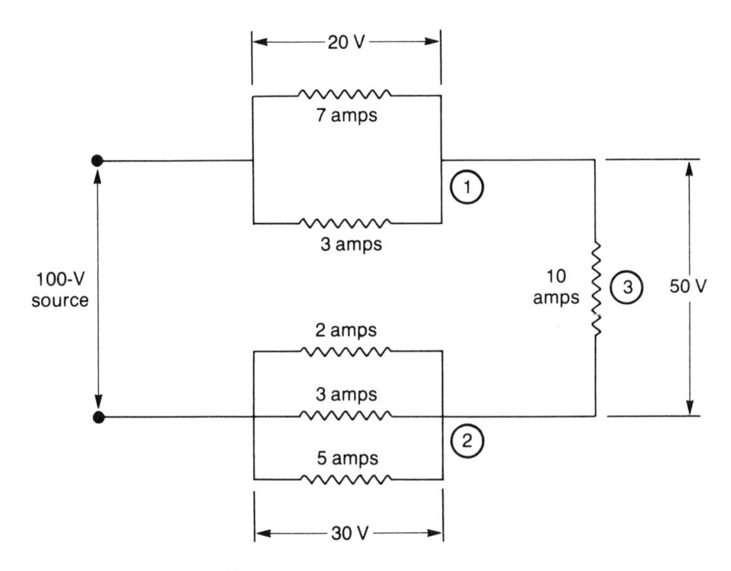

Figure 8–3 Series–parallel circuit.

If resistances are connected in series, the ohms simply add up (as do the volts for generators connected in series), but if they are connected in multiple, it is observed that the resultant resistance is less than the smallest of the component resistances in the circuit. This is so because each additional resistance provides an additional path for the current, so that more current can flow. The conducting ability of the circuit is increased, and the resistance is lowered. It is also observed that connecting in series adds up the ohms (or volts); connecting in parallel adds up the amperes.

Series–Parallel Circuits

An example of resistances connected in series–parallel is given in Figure 8–3. To get the total resistance of this circuit, the resultant resistance of each of the two parallel groups is first determined, then the resistances of groups 1, 2, and 3 are added. The same process applies to any number and type of group.

POWER

When electricity is flowing into an electrical appliance such as a lamp bulb, electrical energy is received by the lamp. The rate of transfer of electrical energy is called electrical power. The unit of electrical power is the watt (after the Scottish inventor James Watt). The watt is too small a unit for many commercial purposes, so the kilowatt is customarily used. A kilowatt is 1000 watts.

Going back to the comparison with the water system, the amount of power developed by the water motor will depend on two factors:

1. The pressure of the water (expressed in pounds per square inch or pounds per square foot)
2. The quantity of water flowing through it per minute (expressed in gallons per minute or pounds of water per minute)

If these two quantities are multiplied together, we obtain *power*, that is, the rate of doing work:

$$\text{power} = \text{pressure} \times \text{quantity of water per minute}$$

In an electric circuit, there is an electrical pressure driving the electrons which corresponds to water pressure forcing water to flow through the pipe; this electrical pressure is measured in volts.

Also, in an electrical circuit, there is an electrical current which is pictured as so many electrons flowing past a point in the circuit each second, and this current is measured in amperes. That is, amperes in an electric circuit corresponds to the quantity of water per minute flowing in a pipe.

Similarly, therefore, in an electric circuit we obtain power by multiplying these quantities together:

$$\text{power} = \text{electrical pressure} \times \text{quantity of electrons per second}$$

$$\text{watts } (W) = \text{volts } (E) \times \text{amperes } (I)$$

By experiment, it is found that there is a relationship between electrical power expressed in watts and mechanical power expressed in horsepower, 1 horsepower being equal to 746 watts:

$$1 \text{ horsepower} = 746 \text{ watts}$$

This simply means that it takes 746 watts of electrical power to do the same work in the same time as one horsepower of mechanical power.

ENERGY

Since power is the rate of expending energy, that is, the energy expended in a given time (seconds, minutes, hours) the total energy expended will be

$$\text{energy} = \text{power} \times \text{time} = E \times I \times T = \text{watts} \times \text{time}$$

where T is the time in seconds, minutes, or hours. The unit of electrical energy most commonly used is the kilowatt-hour. This is what house meters read.

HEAT LOSS

By Ohm's law, $E = IR$. If IR is substituted for E in the expression for power, it becomes

$$W = EI, \quad W = I \times IR$$

This may also be written

$$W = I^2R$$

where I^2, or I squared, signifies I multiplied by itself, that is, $I \times I$.

Therefore, if the current flowing through a circuit is known, as well as its resistance, it is possible to determine the power necessary to overcome the effects of the resistance. Where this power does not produce useful work, that is, where the electrical energy is not converted to some mechanical work, it is converted into heat and dissipated into the surrounding atmosphere. Owing to the electrical resistance encountered, this heat may be likened to the heat developed by friction, and represents a loss.

An example of such a condition is the heat loss in the wires that carry the electric current from the generator to, let us say, an electric motor. In determining R, however, the resistance of the wires of the motor must be included, for there are losses in these as well as in any other wires.

If the current in a wire is *doubled*, the heat loss is therefore *quadrupled* (not doubled) if the resistance remains the same.

In transmitting electric power a great distance, therefore, two factors must be considered. First, the IR drop in the line must not be so great that the electrical pressure or voltage at the receiving end will be insufficient. Second, the I^2R power loss in the line must not be so great that it would be cheaper to carry fuel to the other end of the line and set up a plant there.

REVIEW

- ✦ In every electric circuit there must be a complete path for the flow of current. This path extends from one terminal of a source of electric energy, usually designated as the negative terminal, through the conductor and the device or devices using the energy, through the other or return conductor back to the second or positive terminal.
- ✦ There are two fundamental types of electric circuits: the series circuit and the multiple or parallel circuit. Other types are combinations of these.
- ✦ In a series circuit, all the parts that make up the circuit are connected in series, so that whatever current flows through one of the parts passes through all of the parts.

✦ A multiple or parallel circuit is one in which all of the components receive the full line voltage, the current in each part of the circuit being dependent on the resistance of that part of the circuit to the flow of electricity.

✦ Power is the rate of transfer of energy and is expressed in watts (or kilowatts = 1000 watts). It is the product of voltage and current.

✦ Energy is the expenditure of power over a period of time, and is expressed in watt-hours (1 kilowatt-hour = 1000 watt-hours). It is the product of power and time.

✦ Energy is expended when electricity flows through a resistance, and is converted into heat that is dissipated into the surrounding atmosphere. From Ohm's law, the heat loss can be derived by multiplying the square of the current (in amperes) by the resistance (in ohms).

STUDY QUESTIONS

1. How are electrical appliances connected in series? In parallel?
2. What is the current in each appliance connected in series? In parallel?
3. What is the voltage drop across each appliance connected in series? In parallel?
4. When resistances are connected in series and in parallel, how is the total resistance of the circuit determined?
5. What is a series–parallel connection? How is the total resistance of such circuits determined?
6. What is meant by ''electric power'' in a circuit? In what units is it expressed? What is its relation to horsepower?
7. What two quantities multiplied together give the electric power?
8. Why are energy and power not to be used interchangeably? What is the electrical unit of energy?
9. How is the heat loss in a wire determined? If the current is doubled, what happens to the heat loss?
10. What two factors must be considered in transmitting power a great distance?

chapter 9

Inductance, Capacitance, and Impedance

INDUCTANCE

A conductor carrying alternating current has a magnetic field around it which alternates its characteristics in accordance with the alternations of the current flowing through the conductor (Figure 9–1). Following the "sine-wave" characteristic of the alternating current, the magnetic field about the conductor is zero at the start, builds up to a maximum in one direction at the first quarter cycle (Figure 9–2a, b, c, d), reduces to zero at the second quarter cycle (Figure 9–2d, c, b, a), builds up again to a maximum in the third quarter cycle but in a direction opposite to the previous maximum, and then reduces to zero again to complete the cycle.

Self-Inductance

Now this magnetic field (or magnetic lines of force) acting in this manner constitutes a moving magnetic field which cuts the conductor carrying the current producing it. Such action will produce a voltage in the conductor distinct from that causing the original current to flow. If the right-hand rule for determining the direction of the magnetic field and the right-hand rule for determining the direction of induced voltage are applied, it will be found that the induced voltage

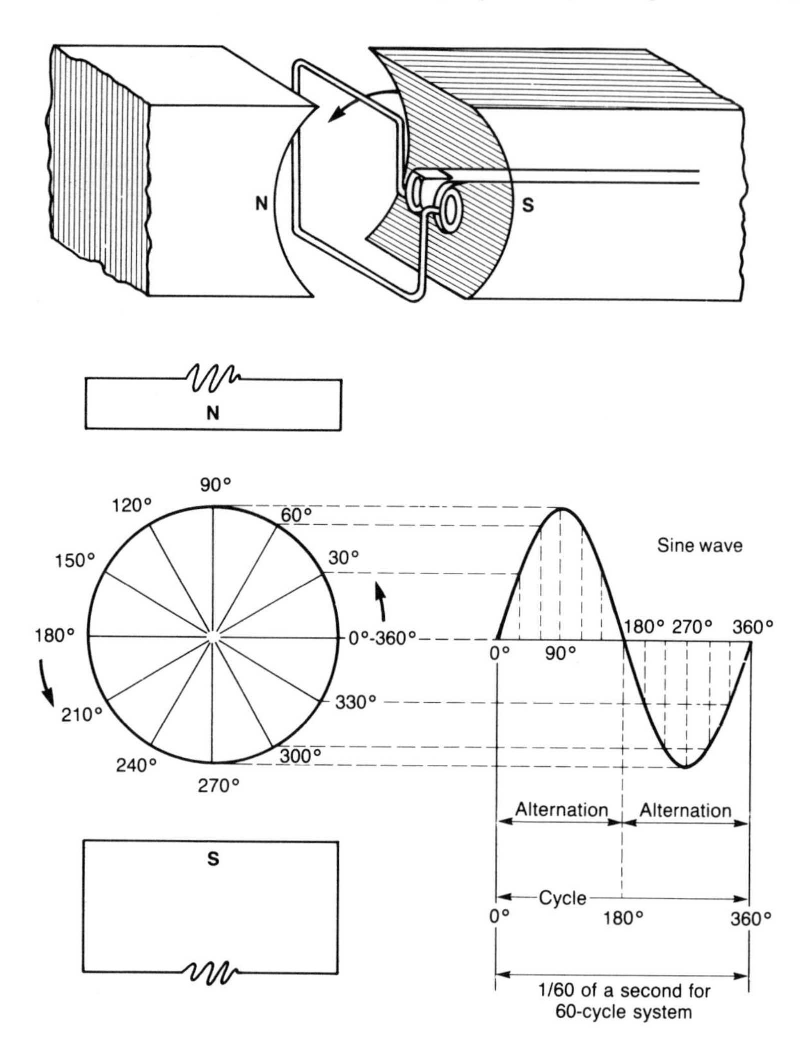

Figure 9–1 Generation of one cycle of alternating current.

will be different from the original voltage in the conductor. Actually, this second voltage also produces a current, which in turn affects the original current, and which in turn affects the magnetic field around the conductor, thus affecting the whole setup. This finally becomes stabilized at some point, and the resultant current flowing in the conductor is taken as a reference. This is the current shown as the "sine wave" in Figure 9–3c.

While the voltage and current in the conductor are reaching their maximum values, the magnetic field about the conductor is reaching its maximum. At the point of maximum, however, the surrounding magnetic field is momentarily

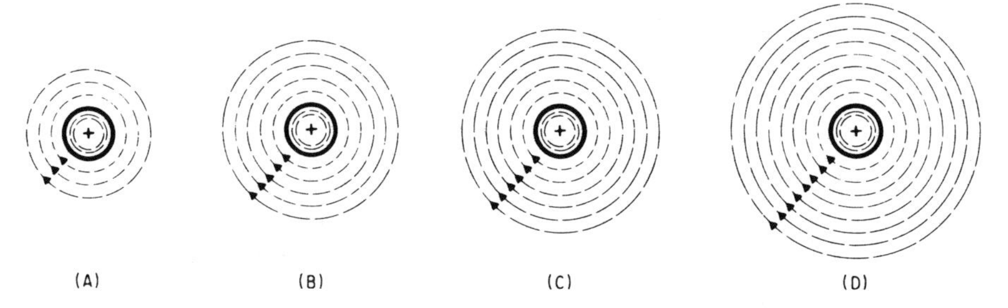

Figure 9–2 (a,b,c,d) Magnetic field expanding about a conductor; (d,c,b,a) Magnetic field contracting about a conductor.

stopped, and hence, at this point, no voltage is being induced in the conductor. Likewise, when the voltage and current are passing from a positive value, through zero, to a negative value, the magnetic field about the conductor is moving most rapidly from a substantial value in one direction and building up to a substantial value in the opposite direction. At this point, therefore, the induced voltage will be a maximum. It is seen, therefore, that the zero and maximum values of the induced voltage do not occur in conjunction with the similar value of the

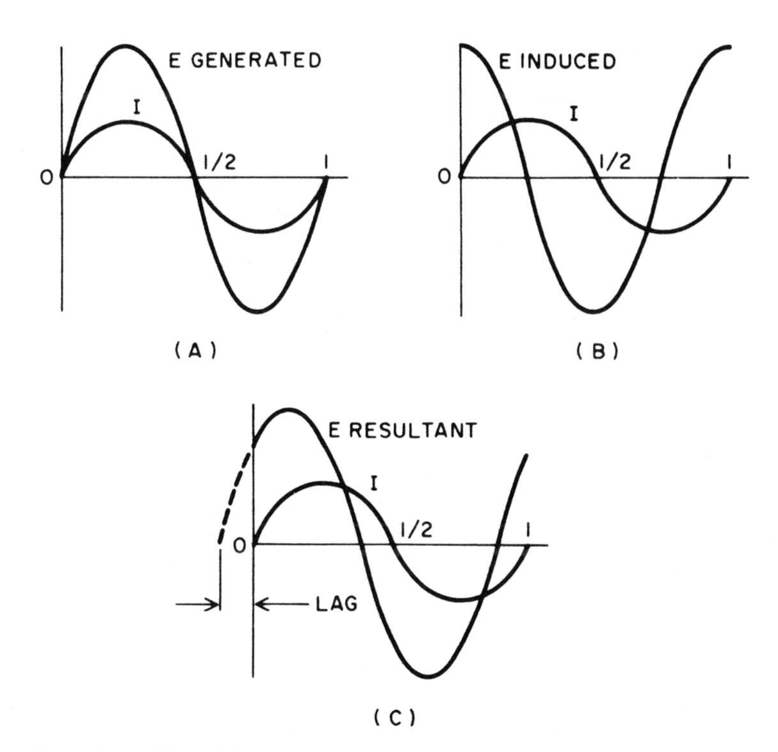

Figure 9–3 Effect of inductance on voltage and current in a conductor (not to scale).

main voltage in the conductor, but are displaced from them by a quarter cycle. This is shown in Figure 9–3b.

If the two voltages in the conductor are now combined or added, the resulting voltage will be as indicated in Figure 9–3c. Actually, of course, only the one resultant voltage exists in the conductor. From Figure 9–3c it is noticed that the voltage wave has been displaced somewhat, so that it reaches its zero and maximum values at some time other than the original voltage. However, the current values still reach their zero and maximum values as they did originally. The net effect of this reaction in the conductor, then, is to cause the current to ''lag'' behind the voltage, as illustrated diagrammatically in Figure 9–3c, where it may be seen that the voltage has reached its maximum and started to fall some time before the current reaches a maximum. Some of the current will be flowing in the circuit at the instant when the voltage is zero. This action resembles that of the heating and cooling of the earth in summer and winter; as is known, the longest days, when the most heat is received from the sun, precede by some weeks the hottest weather, and the coldest weather always occurs sometime after the shortest days when the least heat is received.

Mutual Inductance

The magnetic reaction in a conductor is called self-inductance since it is caused by the magnetic field about itself. This reaction may also be caused by the magnetic fields of adjacent conductors (Figure 9–4), in which case it is called mutual inductance between the conductors, since both conductors affect each other. Self-inductance is sometimes spoken of as electrical inertia, because the electric current resists starting or stopping, increase or lessening, which means that the electrons resist starting or stopping or other change of their motion, just as material bodies do.

Inductive Reactance

An inductive circuit differs greatly from a resistance type of circuit. In the resistance-type circuit, the only obstruction to the flow of electricity is that due to the physical resistance of the circuit. In a circuit having inductance but no resistance, the only obstruction to the flow of electricity is that which is due to the voltage generated or induced by the circuit itself. (Strictly speaking, a circuit having no resistance cannot be built.) Such an obstruction is called the *reactance* of the circuit. Since the reactance is treated as an obstruction to the flow of electricity, the unit by which values of reactance are expressed is the ohm. To avoid confusion in practice, the letter R is commonly used to denote resistance, the letter X to denote reactance (more specifically, X_L to denote reactance due to inductance).

Assume that a voltage of 100 volts is applied across a circuit whose resistance is 2 ohms. Then, by Ohm's law,

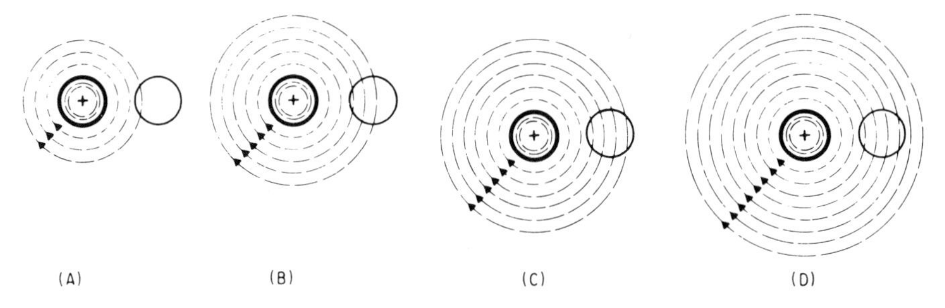

Figure 9–4 Illustrating the effect of the magnetic field about a conductor on an adjacent conductor.

$$I = \frac{E}{R} = \frac{100}{2} = 50 \text{ amperes}$$

The voltage and current relations are the same as in Figure 9–3a.

Assume that the same voltage of 100 V is applied across a circuit which has no resistance but presents an obstruction to the flow of electricity caused by inductance which is equivalent to 2 Ω, the same amount as that of the resistance circuit above. Then, by Ohm's law,

$$t = \frac{E}{X_L} = \frac{100}{2} = 50 \text{ amperes}$$

The voltage and current relations are the same as in Figure 9–3b. Note that when there is only inductance obstructing the flow of electricity, the current wave "lags" the voltage wave by a quarter cycle.

The inductance of a circuit or piece of equipment is dependent on its physical characteristics. The unit of measurement of inductance is called the "henry" (after the American physicist Joseph Henry) and is usually expressed by the symbol L. The effect of this inductance on the flow of electricity is called reactance—or inductive reactance—and is measured in ohms.

RESISTANCE AND INDUCTANCE

Now, assume that both the resistance and inductance are combined in one circuit with the same applied voltage of 100 volts. The current flowing in this circuit when measured is found to be approximately 35.5 amperes. By Ohm's law,

$$100 \text{ volts} = 35.5 \text{ amperes} \times 2.82 \text{ ohms}$$

Thus it is seen that the resultant obstruction to the flow of electricity in this case is greater than any one of the factors (2 ohms), but is not equal to the arithmetic sum of both factors (4 ohms). In the first instance it is obvious that the inductance of the circuit is an added obstruction and that the resultant obstruction is bound to be greater than 2 ohms. In the latter instance, referring to Figure 9–3a and b, if the voltage waves are made coincident, it will be noticed that the currents are not coincident, and therefore the effects of resistance and inductance on the current are not coincident. By analysis it is found that the resultant obstruction may be found by obtaining the square root of the sum of the squared values of resistance and inductive reactance; that is,

$$Z = \sqrt{R^2 + X_L^2}$$

$$= \sqrt{(2)^2 + (2)^2} = \sqrt{8} = 2.82 \text{ ohms}$$

where Z is the resultant obstruction and is called *impedance* to distinguish it from the other components.

It will be noted from Figure 9–3c, that when resistance is combined with inductance, the current wave lags the voltage wave less than in Figure 9–3b when only inductance existed in the circuit. The more resistance added, the closer will the current and voltage waves approach each other.

CAPACITANCE

When two conductors are in the proximity of each other, there is yet another reaction between them, but it is not due to the magnetic field. When an alternating current flows through a conductor, it becomes alternately positively and negatively charged, as described in Chapter 7. That is, during one half-cycle of the alternation, there will be a scarcity of electrons, while in the next half-cycle there will be an excess of electrons. During the first half-cycle, the conductor having a scarcity of electrons will attempt to pull some away from the adjacent conductor, even though there may be air or other insulating material between the conductors. The result will be that a number of additional electrons will be pulled into the second conductor due to the attraction of the first conductor. Now, during the second half-cycle, the exact opposite happens, that is, electrons will be pushed out of the second conductor because of the effect of the first conductor. Thus it is seen that a to-and-fro, or alternating, circulation of electrons is set up in the second conductor; in other words, an alternating current is set up in the conductor.

If both conductors are carrying alternating currents, they will therefore react upon each other in the manner described above. The amount of this reaction, called *capacitance*, will depend on the areas of the conductors exposed to each other, the distance and type of insulating material between them, and the number of electrons active in each conductor during the alternations (which may be said to be current or voltage in the conductors). Like the inductance in a circuit, the current set up in a conductor because of this "capacitance" effect will be displaced from the "normal" current by a quarter of a cycle, as it is evident that when the normal current is positive, the "capacitance" current will be

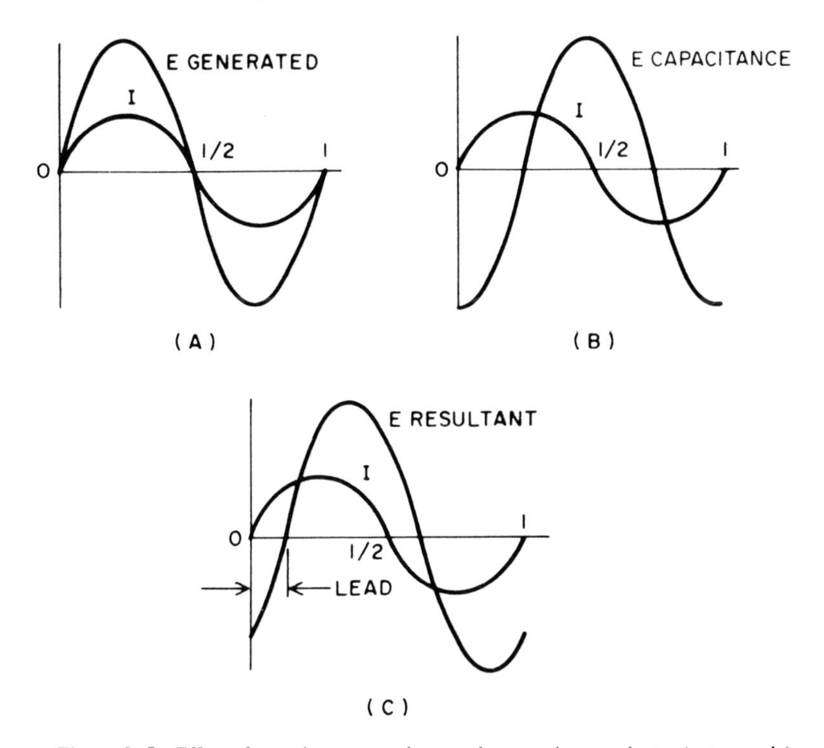

Figure 9–5 Effect of capacitance on voltage and current in a conductor (not to scale).

negative; this time, however, the current wave will "lead" or be ahead of the voltage wave by a quarter-cycle. This is shown in Figure 9–5b.

Water Analogy of A Capacitor

Refer to Figure 9–6: As the pump moves to the right, water flows in the top pipe to the cylinder at the right. The flexible diaphragm in the cylinder will move to the right (as shown), transmitting its motion to the water to its right during the motion of the pump. In the leftward motion of the pump, the reverse action takes place. The net result is that an alternating-current flow is set up in both the pipes, without a direct connection between the water in the lower pipe and that in the upper pipe.

In the process, because of the flexibility of the diaphragm, the current of water will tend to arrive at the undistended diaphragm before the pump completes its stroke in one direction. This may be observed by the relative positions of the driving motor on the left and the water in the driven machine on the right. Compared to the flow of water in the pipes driving a water motor shown in Figure 7–3c, the current flow may be said to "lead," or act ahead of, the pressure being applied by the pump at the left. The mechanical stresses on the

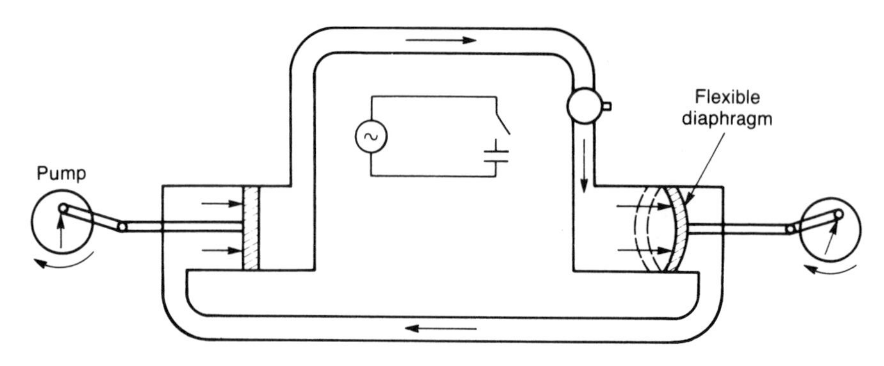

Figure 9–6 Water analogy of a capacitor in an alternating-current circuit.

diaphragm, alternately compression on one side and elongation on the other, correspond to the electrostatic stresses on the dielectric in the capacitor.

CAPACITIVE REACTANCE

As with inductive circuits, such an obstruction of a condenser circuit is called the reactance of the circuit. It is sometimes called capacitive reactance to distinguish it from inductive reactance. Since it is also an obstruction to the flow of current, it is measured in ohms and the symbol X_C is used to denote the reactance due to capacitance.

The capacitance of a circuit or piece of equipment is also dependent on its physical characteristics. The unit of measurement of capacitance is called the "farad" (after the English scientist Michael Faraday) and is usually expressed by the symbol C. The effect of this capacitance on the flow of electricity is also called reactance—or capacitive reactance—and is measured in ohms.

RESISTANCE AND CAPACITANCE

If we compare currents flowing through a circuit having only resistance (2 Ω), a circuit having only capacitance (also equivalent to 2 Ω), and a circuit having a combination of the two, the same results will be obtained as were obtained previously; that is:

$$I = \frac{E}{R} = \frac{100}{2} = 50 \text{ amps}$$

$$I = \frac{E}{X_C} = \frac{100}{2} = 50 \text{ amps}$$

$$I = \frac{E}{X} = \frac{100}{2.82} = 35.5 \text{ amps}$$

In this case, too, $Z = \sqrt{R^2 + X_C^2}$. It will be noted from Figure 9–5c that the current wave now "leads" the voltage wave and that the amount of "lead" is less than in Figure 9–5b. In this case, too, the more resistance added, the closer will the current and voltage waves approach each other.

IMPEDANCE

The total obstruction to the flow of current in a circuit may, therefore, be caused by resistance, inductance, and capacitance—or, more generally, by resistance and reactance. This total obstruction (which impedes the flow of electricity) is called the impedance of the circuit and, as mentioned previously, is usually denoted by the letter Z. In alternating currents, Ohm's law becomes

$$I = \frac{E}{Z} \quad \text{or} \quad E = IZ \quad \text{or} \quad Z = \frac{E}{I}$$

RESISTANCE, INDUCTANCE, AND CAPACITANCE

It will be observed from Figures 9–3b and 9–5b that the currents due to inductance and capacitance are in direct opposition to each other. Thus, if these two circuits were combined, the currents would exactly balance each other and the resultant current would be zero. Going one step further, if the three circuits (resistance only, inductance only, and capacitance only) were combined, the resultant current would be that due to the resistance alone. It is evident, then, that the effects of inductance and capacitance tend to nullify each other. The net reactance effect, then, can be said to be the difference between the inductive and capacitive reactances, that is,

$$X = X_L - X_C$$

It therefore follows that the resultant impedance of the circuit will be

$$Z = \sqrt{R^2 + X^2}$$

or, substituting,

$$Z = \sqrt{R^2 + (X_L - X_C)^2}$$

The current flowing in such a circuit will then be, by Ohm's law,

$$I = \frac{E}{Z} = \frac{E}{\sqrt{R^2 + (X_L - X_C)^2}}$$

RESONANCE

Here again it is evident that if X_L is equal to X_C, then $X_L - X_C$ is equal to zero and the only quantity left will.be R, the resistance. When this condition occurs in practice, the circuit is said to be in *resonance*. The relative position of the current wave with respect to the voltage wave will be determined by the relative values of resistance, inductance, and capacitance of the circuit.

REVIEW

- A conductor carrying alternating current has a magnetic field around it which alternates its characteristics in accordance with the alternations of the current flowing through the conductor.
- The alternating magnetic field cuts the conductor itself inducing in it a voltage that tends to oppose and distort the voltage in the conductor. This is known as self-inductance.
- The effect of these two voltages, actually one resultant voltage, is, in effect, to have the voltage lag the current flowing in the conductor.
- The alternating magnetic field around a conductor cuts an adjacent conductor, inducing in it a voltage. The second conductor, carrying an alternating current, has the same effect on the first conductor. This is known as mutual inductance.
- The effect of these inductances is to cause an obstruction to the flow of current, but essentially at right angles to the obstruction caused by the resistance of the conductor. The net effect is expressed in ohms.
- Similar effects in the conductors are also caused by the capacitance effect of the conductors. The net effect is to obstruct the flow of current, but essentially at right angles to the obstruction caused by the resistance of the conductors. The net effect is also expressed in ohms.
- The inductance effect and the capacitance effect, while both at right angles to the resistance effect, are in opposition to each other. The net effect of resistance, inductance, and capacitance is known as impedance, and is measured in ohms.
- The condition where the inductance effect and the capacitance effect cancel each other, leaving only the resistance effect, is known as resonance.

STUDY QUESTIONS

1. What is inductance? What is the difference between self- and mutual inductance?

2. What is the effect of inductance on the relation between voltage and current in a circuit?
3. What is the difference between inductance and inductive reactance? In what units are they expressed?
4. What is capacitance?
5. What is the effect of capacitance on the relation between voltage and current in a circuit?
6. What is the difference between capacitance and capacitive reactance? In what units are they expressed?
7. What is impedance?
8. How is Ohm's law affected by alternating current?
9. What is the difference between resistance, reactance, and impedance? What is the relationship between them?
10. What is resonance, and what is its effect in a circuit?

chapter 10

Transformers and Autotransformers

GENERATION OF VOLTAGE IN A TRANSFORMER COIL

In Chapter 9 it has been demonstrated how an electrical pressure—or voltage—may be generated in a conductor adjacent to another one carrying an alternating current. This process is called induction, and in the conductors it acted to obstruct the normal flow of current. This same effect of induction, however, can prove useful in another way.

Instead of two wires adjacent to each other, assume that there are two coils of wire adjacent to each other and that an alternating current of electricity flows in one of them (Figure 10–1). As was shown previously, the voltage induced in the second coil will depend on the length of the conductor, the relative speed between conductor and the magnetic field, and the strength of the magnetic field. The entire magnetic field set up by the first coil can be assumed to cut the turns of the second coil if both coils are wound on an iron core. The relative speed between the conductor and magnetic field is fixed, being dependent on the frequency of the alternating current flowing through the first coil. The voltage in the second coil will therefore depend on the length of the conductor—or on the number of turns. If the magnetic field and the rate of cutting are the same for both coils, each turn of each coil will have the same voltage produced in it. Therefore, to obtain the desired voltage in the second coil, the volts per turn

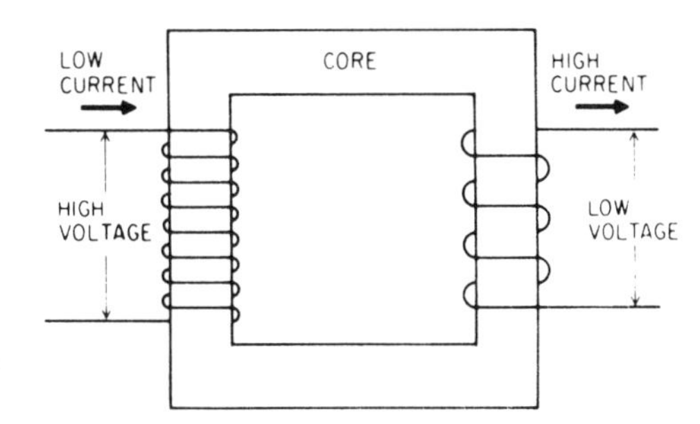

Figure 10–1 Diagram of transformer having two windings insulated from each other and wound on a common iron core.

are determined from the first coil and the proper number of turns wound on the second coil. If a voltage of 1000 volts is applied to a coil of 1000 turns, 1000 volts will be generated in the entire coil, each turn generating one volt. Now, if only a voltage of 100 volts is desired in the second coil, only 100 turns will be required as each turn generates one volt.

It must be remembered that the voltage in the second coil is an induced voltage and will therefore be displaced from the voltage in the first coil by a half-cycle (Figure 10–2). The currents will also be displaced by a half cycle.

RATIO OF TRANSFORMATION

The usual type of power transformer consists of two electrical windings placed on a common iron core, the number of turns in each winding being dependent on the desired ratio of transformation as described above. One winding, which is called the *primary winding*, is connected to the source from which power is supplied. The circuit to which power is delivered is connected to the other or *secondary winding* of the transformer. The primary and secondary windings are insulated from each other and from the iron core. Hence the transformer not only is a device by which the voltage of an alternating-current system may be changed, but it also serves as a safety device for effectively isolating the lower voltage circuit from the higher-voltage circuit.

The magnetic field produced by the first coil is determined by the number of amperes flowing in the turns of the coil, that is, on the ampere-turns. Since

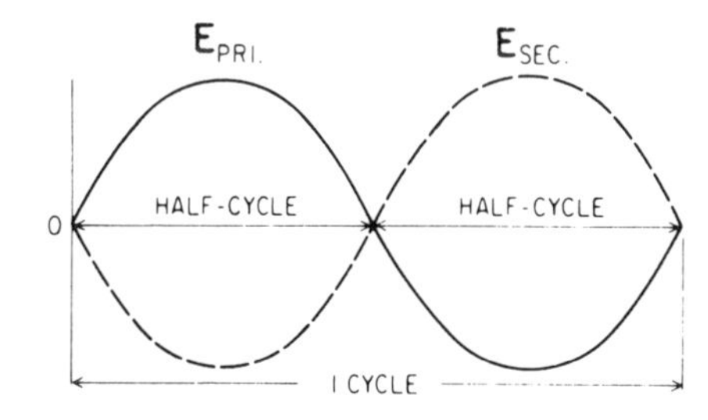

Figure 10–2 Curves showing voltage in secondary of transformer displaced from voltage in primary by one-half cycle.

the same magnetic field is responsible for the current induced in the second coil, the ampere-turns of the second coil must match the ampere-turns of the first coil. Therefore, in the coils mentioned above, the current in the first coil will be one-tenth that of the second; for instance, if the current in the first coil is 1 ampere, the current in the secondary winding (or second coil) will be 10 amperes.

Another way to prove the same thing is to measure the power input and compare it to the output. Remembering from Chapter 8 that power is represented by the product of the voltage and current, then, neglecting losses,

$$I_{\text{pri}} \times E_{\text{pri}} = I_{\text{sec}} \times E_{\text{sec}}$$

or

$$1 \text{ amp} \times 1000 \text{ volts} = 10 \text{ amps} \times 100 \text{ volts}$$

All of the relations above may be expressed simply as follows:

$$\frac{N_P}{N_S} = \frac{E_P}{E_S} = \frac{I_S}{I_P}$$

where N_P = number of turns in primary winding
N_S = number of turns in secondary winding
E_P = voltage induced in primary winding
E_S = voltage induced in secondary winding
I_P = current flowing in primary winding
I_S = current flowing in secondary winding

ACTION OF TRANSFORMER UNDER LOAD

The transformer also acts automatically to regulate the flow of energy in the primary as it is demanded by the load connected to the secondary. When no circuit is connected to the secondary coil, no current will flow in it. However, a current will flow in the primary coil because it presents a continuous path. This current will set up an alternating magnetic field which will cause a voltage to be self-induced in the primary coil. If there are no losses considered in the transformer, that is, if a "perfect" transformer is assumed, the self-induced voltage will exactly balance the original voltage in the primary coil; the net result will be a zero voltage and consequently, a zero current.

Let a load be connected to the secondary:

1. A continuous circuit will be established and current will therefore flow in the secondary coil.
2. This secondary current in turn creates an alternating magnetic field about

it which will tend to induce within itself a voltage of such value as to prevent the current from flowing.

3. At the same time, this secondary magnetic field will act on the primary in such a way as to reduce the effect of self-induction in the primary coil.
4. This will allow more current to flow in the primary coil, which in turn will set up a stronger magnetic field.
5. This stronger magnetic field will induce a greater voltage in the secondary coil, which will in turn tend to reduce the effect of self-induction in the second coil.

Thus a state of equalization is reached where the magnetic fields of the primary and secondary are so balanced that the ampere-turns in one match the ampere-turns in the other.

As more load is connected to the secondary coil, more paths will be provided for the current and flow. The impedance of the circuit connected to the secondary will be reduced and consequently more current will flow. The cycle of events mentioned previously will then recur until a new state of equalization is reached. This new balance, it is obvious, can be obtained only by increasing the primary current. The same chain of events is repeated in the reverse sequence when the load connected to the secondary coil is reduced.

TRANSFORMER LOSSES

All that has been said refers to an ideal or perfect transformer. In any actual instrument, the output is less than the input because of the *transformer losses*—heat developed in the wires and in the core.

Copper Losses

The heat losses developed in the wires are those caused by the resistance to the flow of current, which was likened in Chapter 7 to the friction of water in a pipe, or friction between electrons moving in the conductor. This loss is determined by multiplying the square of the value of the current by the resistance, that is I^2R. It must be emphasized that only the resistance R causes this effect. The inductance, L, does not resist the flow of current; it prevents it from coming into being. Therefore, in determining the heat losses, only the resistance, R, is to be considered; the inductance L (or X_L) and the impedance Z are not to be considered. It is obvious, too, that the greater the current flowing in the transformer (whether primary or secondary), the greater will be the heat or I^2R losses in the conductors of the coils. These losses are usually referred to as the *copper losses* in a transformer.

Hysteresis

If each molecule of the iron comprising the core of a transformer is considered as a minute magnet, then, as the magnetic field set up by an alternating current changes direction, these small magnets will reverse their position to accommodate the strong magnetic field. As these molecules change their position with each alternation of the magnetic field, friction between them is produced and energy is used up in overcoming it. This loss of energy due to friction between the molecules is given off as heat (and is called *hysteresis*).

Eddy Currents

The alternating magnetic field set up in a transformer not only induces voltages and currents in the coils through which it passes, but does so also in the core of iron. The currents thus induced in the iron core swirl around like eddies in a pool of water, and are therefore appropriately named *eddy currents*. These currents in the core also produce an I^2R heat loss.

Iron Losses

The sum of both the hysteresis and eddy current losses is usually referred to as the *iron losses* in a transformer. Since there is only a relatively small difference in the magnetic field in the iron core at any time, as explained above, these losses will present little variation as the load on the transformer is increased or decreased.

To hold down the eddy current losses in the core of a transformer, it is usually built up of leaves or laminations of iron. The idea behind this is that the various laminations will have a certain amount of resistance between them and the currents set up in them will not add together as they would in a solid piece of iron. Thus the value of the current is reduced and the I^2R loss in the iron core due to these eddy currents is also reduced.

No-Load Losses

It will be observed that when the secondary of a transformer is open, that is, it has no load connected to it, a small current will nevertheless flow in the primary. As mentioned previously, the large self-induced voltage in the primary practically counterbalances the original voltage in that coil. The current that the primary takes under these circumstances (sometimes referred to as the *exciting current*) serves to supply the hysteresis and eddy current (or iron) losses as well as a small I^2R loss in the primary coil itself. The sum of these losses is usually referred to as the *no-load loss* in a transformer.

METHODS OF COOLING

The heat generated in transformers due to the unpreventable iron losses and copper losses must be carried away to prevent excessive rise of temperature and injury to the insulation about the conductors. The cooling method used must not only maintain a sufficiently low average temperature but must prevent an excessive temperature rise in any portion of the transformer; that is, it must prevent "hot spots." For this reason the core and coils of the transformer are usually submerged in oil that is free to circulate. Air or some special fluids (askarels) are also used for special applications.

TRANSFORMER RATINGS

The capacity of a transformer is limited by the permissible temperature rise. Both the current and the voltage contribute to determine the heat generated in a transformer. For this reason, the rating of a transformer is expressed as a product of the volts and amperes or in volt-amperes. For practical purposes, the kilovolt-ampere (kVA) is used.

EFFICIENCY

Like any other piece of equipment, the efficiency of a transformer is given by

$$\frac{\text{output}}{\text{input}} = \frac{\text{watt-hours output}}{\text{watt-hours input}}$$

For commercial transformers, this efficiency at full rated load is in the nature of 98 or 99%.

AUTOTRANSFORMERS

When the ratio of transformation desired is low, in the nature of 2 or 3 to 1, and the isolating feature of the two-winding transformer is not essential, use is made of the autotransformer. An autotransformer (sometimes also known as a *compensator*) is a transformer that has only one winding, a portion of which serves both as primary and secondary (Figure 10–3). In this type of transformer, a portion of the electrical energy is transformed and the remainder flows conductively through its windings.

Figure 10–3 shows a schematic diagram of an autotransformer. If a source of voltage E_P, is applied across all the turns between a and c, this part of the coil will serve as a primary winding. Some of the turns, between b and c, will

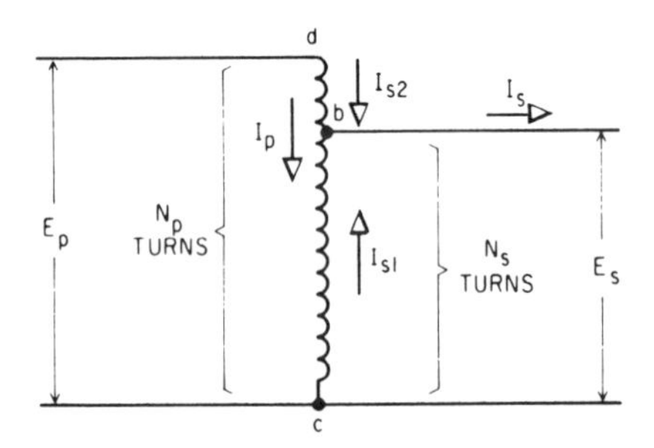

Figure 10–3 Schematic diagram of an autotransformer.

also serve as a secondary winding giving the voltage E_S. This arrangement uses the autotransformer for step-down purposes; reversing E_P and E_S will allow the transformer to be used equally well as a step-up transformer. In this transformer, the same ratio of transformation holds as in the two-winding transformer:

$$\frac{N_P}{N_S} = \frac{E_P}{E_S} = \frac{I_S}{I_P}$$

When the primary current I_P flows in the direction shown by the arrow in the figure, the secondary current I_S will flow in the opposite direction. Hence, in the portion of the windings between b and c, the current is the difference between I_P and I_{SI}. When the ratio of transformation is small, the difference between E_P and E_S is small and also the difference between I_P and I_{SI}, so that the portion of the winding between b and c, which carries the difference of these currents, can be made of a small cross-section conductor since it will have to carry only a small current.

Under these circumstances, the autotransformer is very much cheaper than the two-coil transformer of the same kilovolt-ampere rating. The autotransformer has the disadvantage, however, that the primary and secondary circuits are electrically connected and therefore cannot safely be used for stepping down from a high voltage: for example, 2400 volts, to a voltage suitable for lamps or motors, that is, 120/240 volts.

REVIEW

- The effect of an alternating magnetic field produced by one conductor upon an adjacent conductor is made use of in the transformer. The two single conductors are replaced by two coils, insulated from each other and wound on a common steel core.

- The strength of the magnetic field in the incoming winding (called the primary) is determined by the number of turns in the winding and the current flowing in it, that is, the ampere-turns. Neglecting losses, the same ampere-turns cause an effect in the second or secondary winding; the current flowing in the secondary will depend on its number of turns, so that the ampere-turns in both primary and secondary are the same.

✦ The voltage generated in the secondary will depend on the number of turns. The ratio of the input or supply voltage to the output or secondary voltage will be the same as the ratio of the number of turns in the two windings.

✦ Neglecting losses, the power input should be the same as the power output. Hence the product of the voltage and current in the primary should be the same as that of the secondary voltage and current. The effect is that the currents in the primary and secondary will be inversely proportional as the ratio of transformation.

✦ Losses in the transformer include those due to the resistance of the conductors in both primary and secondary windings, known as copper losses. Losses also occur from eddy current and hysteresis losses in the steel core. These are known as iron losses.

✦ When no load is applied to the secondary of a transformer, a small current will nevertheless flow, known as the exciting current; this supplies the copper losses in the primary and the iron losses in the core.

✦ The heat created by the currents flowing in the windings is dissipated to the atmosphere either directly in air-cooled transformers, or through oil surrounding the coils in oil-cooled units.

✦ Transformers are rated by the voltage and current limitations, and are expressed in volt-amperes or kilovolt-amperes.

✦ Efficiency of transformers is very high, in the range of 98 to 99%.

✦ When the ratio of transformation desired is low, the two windings may be combined into one, and the unit called an autotransformer. Here a portion of the energy is transformed and the remainder flows conductively through the winding.

STUDY QUESTIONS

1. What is a transformer? Mention two purposes for which transformers are used.
2. How is a voltage generated in a coil of the transformer?
3. What is the phase relation of the primary and secondary voltages?
4. What is the ratio of transformation?
5. Describe what takes place in a transformer when load is connected to the secondary.
6. What are the transformer losses? What form do these losses take?
7. In a transformer, what are copper losses; hysteresis; eddy currents; iron losses; no-load losses?

8. Why must transformers be cooled? Mention several methods used for cooling transformers.
9. How are transformers rated?
10. What is an autotransformer? What are some of its advantages and disadvantages?

chapter 11

Electrical Measurements

COMPARATIVE VALUES OF AC AND DC VOLTAGE AND CURRENT

The concepts of electrical pressure, current, and resistance or impedance, and the relations existing between them in Ohm's law, have been explained previously. In direct-current circuits, voltage and current can readily be evaluated, a volt being a fixed continuous pressure which causes the electrons to circulate, and an ampere being so many electrons flowing past a given point in a given time. In alternating currents, these quantities cannot be evaluated so readily. Consider the evaluation of voltage for one cycle: During the first half-cycle, it has varying values in a positive direction, while in the second half-cycle, the values reverse in a negative direction; the net result of both half-cycles is therefore zero. A similar situation exists with respect to the current values. It is apparent, therefore, that a different method of evaluating current and voltage in an alternating-current circuit must be conceived.

EFFECTIVE VALUES OF VOLTAGE AND CURRENT

When an electric current flows through a circuit, heat is generated or produced in the circuit—as demonstrated in Chapter 8. The rate at which heat is produced in the circuit is equal to I^2R, where I is the current in amperes and R the resistance in ohms. This heating effect takes place when either a direct (nonalternating) or an alternating current flows.

In an alternating-current circuit, however, the rate at which heat is being generated is constantly changing because the current is constantly changing. In a direct or nonalternating current circuit, when a current flows, the rate at which heat is developed is constant (for a certain amount of current) because the direct current remains constant in amount. For example, an alternating current which reached a maximum value of 100 amperes would not have the same heating effect as a direct current of 100 amperes. To produce the same heating effect, the alternating current would have to be of such amount that it reached a maximum value of 141.4 amperes. Such a current varying according to a sine wave between zero amperes and a maximum of 141.4 amperes has the same effective heating value as a direct (constant) current of 100 amperes and hence its "effective" value is said to be 100 amperes.

The same effective value is used for voltages. Thus an alternating (sine wave) voltage varying between zero volts and 141.4 volts is said to have an "effective" value of 100 volts.

Both alternating currents and voltages are usually expressed in terms of effective values. The ratio between the effective and maximum values is illustrated in Figure 11–1. This ratio can be expressed in either of two ways:

$$\frac{\text{effective value}}{\text{maximum value}} = \frac{0.707 \times \text{maximum value}}{1.414 \times \text{effective value}}$$

Thus, if one value is known, the other can easily be determined.

The maximum value is an instantaneous value. Any other instantaneous values for other instants during the cycle can be determined by scaling off values from the curve of a sine wave. Alternating-current instruments, such as ammeters and voltmeters, are calibrated to indicate the effective values.

POWER

Power, or the rate at which electrical energy is transformed into heat or mechanical energy, is equal to the product of the voltage and the current, as

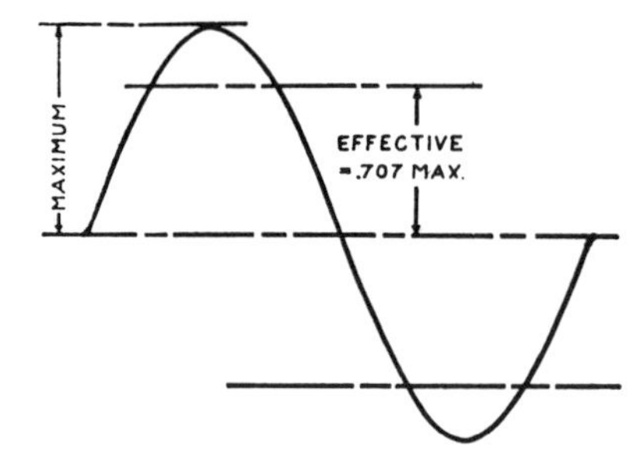

Figure 11–1 Ratio between maximum and effective values of sine-wave voltage or current.

explained in Chapter 8. In an alternating-current circuit, the power (rate of change of energy) at any instant is equal to the product of the voltage and current at that instant. In Table 11–1 are listed the values of the voltage, the current, and the power for a few instants of time during the cycle for the resistance, inductance, and capacitance circuits illustrated in Chapter 9. It will be remembered that a voltage of 100 volts was applied across a resistance, inductive reactance, and capacitive reactance, each having a value of 2 ohms.

Curves showing the variation in power for a complete cycle for these conditions are given in Figure 11–2. Also shown are the cycle of current and voltage values. Diagrams of the circuits are included for reference.

Power in Resistance Circuits

From the tabulation (and the curves plotted) it is seen that in a resistance circuit, the power varies throughout the cycle, increasing from zero to a maximum (of 9997 watts) and then decreasing to zero again twice in each cycle.

It is important to note that at all times during the cycle (with the exception of the brief instants when there is no power), the direction of power, or the change of energy, is always the same. The change is always from electrical energy to other forms of energy. There is no return at any time during the cycle from the other forms of energy back to the electrical form. This is an important characteristic of the resistance circuit. It is the only type of circuit in which all of the electrical energy delivered to the terminals of the circuit is converted into other forms of energy with no return of electrical energy to the source that supplies it.

Power in Inductive Circuits

When an electric current flows in an inductive circuit, electrical energy is changed to another form that is stored up in the magnetic field of the circuit. This stored-up energy in this type of circuit is always changed back to electrical energy. While the current that flows in an inductive circuit is increasing, energy is being stored up in the magnetic field; while the current is decreasing, the magnetic energy is being returned back into the electrical form. Thus in the inductive circuit, there occurs a transfer of energy in two directions, from electric to magnetic, and vice versa, while in the resistance-type circuit there is a transfer of energy in one direction only, from electrical energy to other forms of energy.

The power curve in Figure 11–2b shows the rate at which energy is changing from the electric to the magnetic form, or vice versa. The values of power for those instants when the change of energy is from the electric to the magnetic form (while the current, I, is increasing in amount) are plotted above the baseline. The values of power for those instants when the change of energy is in the reverse direction from the magnetic to the electric form (while the current is decreasing in amount) are plotted below the baseline.

TABLE 11–1 Values of Voltage, Current, and Power

	Resistance Circuit			Inductive Circuit			Capacitive Circuit		
Instant	Voltage ×	Current =	Power	Voltage ×	Current =	Power	Voltage ×	Current =	Power
0°	0.0 ×	0.0 =	0.0	0.0 ×	−70.7 =	0.0	0.0 ×	70.7 =	0.0
30°	70.7 ×	35.3 =	2495.7	70.7 ×	−61.2 =	−4326.8	70.7 ×	61.2 =	4326.8
60°	122.5 ×	61.2 =	7497.0	122.5 ×	−35.3 =	−4324.3	122.5 ×	35.3 =	4324.3
90°	141.4 ×	70.7 =	9997.0	141.4 ×	0.0 =	0.0	141.4 ×	0.0 =	0.0
120°	122.5 ×	61.2 =	7497.0	122.5 ×	35.3 =	4324.3	122.5 ×	−35.3 =	−4324.3
150°	70.7 ×	35.3 =	2495.7	70.7 ×	61.2 =	4326.8	70.7 ×	−61.2 =	−4326.8
180°	0.0 ×	0.0 =	0.0	0.0 ×	70.7 =	0.0	0.0 ×	−70.7 =	0.0
210°	−70.7 ×	−35.3 =	2495.7	−70.7 ×	61.2 =	−4326.8	−70.7 ×	−61.2 =	4326.8
240°	−122.5 ×	−61.2 =	7495.0	−122.5 ×	35.3 =	−4324.3	−122.5 ×	−35.3 =	4324.3
270°	−141.4 ×	−70.7 =	9997.0	−141.4 ×	0.0 =	0.0	−141.4 ×	0.0 =	0.0
300°	−122.5 ×	−61.2 =	7495.0	−122.5 ×	−35.3 =	4324.3	−122.5 ×	35.3 =	−4324.3
330°	−70.7 ×	−35.3 =	2495.7	−70.7 ×	−61.2 =	4326.8	−70.7 ×	61.2 =	−4326.8
360°	0.0 ×	0.0 =	0.0	0.0 ×	−70.7 =	0.0	0.0 ×	70.7 =	0.0
Average power		=	5750.0		=	0.0		=	0.0

Note: Two negative values when multiplied by each other yield a positive result.

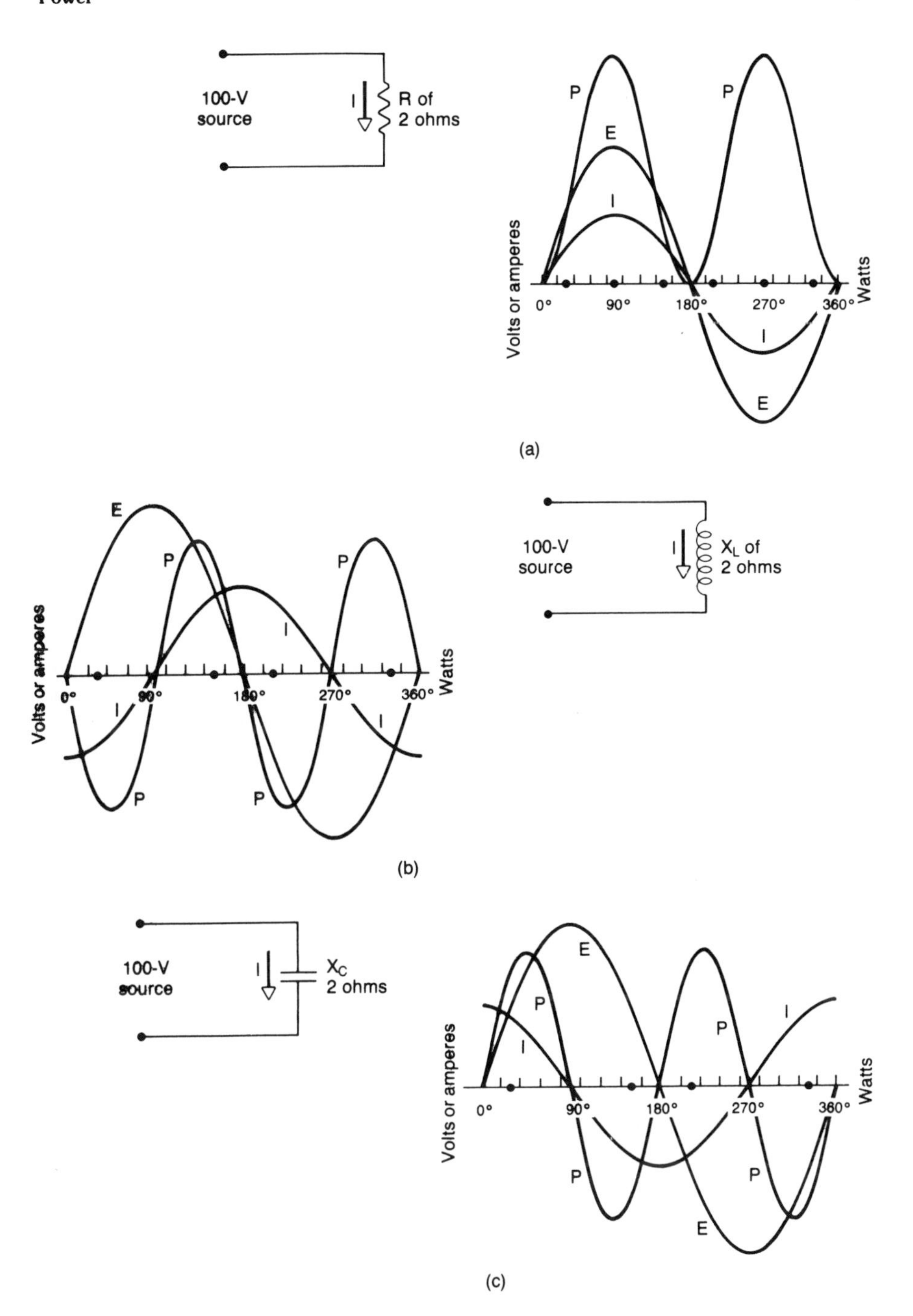

Figure 11–2 Power values in (a) resistance, (b) inductance, and (c) capacitive circuits.

A study of the tabulated values and of the power curve in Figure 11–2b shows that there are two complete cycles of power during each voltage cycle. Furthermore, the amount of energy which is changed from the electric to the magnetic form during the second and fourth quarters of the cycle illustrated is exactly equal to the amount returned from the magnetic to the electric form. Hence the net change of energy or power is zero. All of the energy that is delivered from the generator to the inductive circuit is returned to the generator (it is assumed there is no resistance in the circuit).

Power in Capacitive Circuits

In a capacitive circuit, electrical energy is changed into electrostatic energy and then back to electrical energy in much the same manner as in an inductive circuit. From the tabulation and curves plotted, it is seen that as in the case of a purely inductive circuit, all of the electrical energy delivered to the circuit is returned and the net change of energy or power is zero.

POWER FACTOR

Since the net transfer of energy from the source to the circuit, in both inductive and capacitive circuits, is zero, the product of voltage and current does not indicate the true power of the circuit. Such a product is usually termed the *apparent power* and is expressed in volt-amperes instead of watts. To calculate the true power, the apparent power is multiplied by a factor called the *power factor*.

The power factor of a circuit is the ratio between the true and the apparent power; that is,

$$\text{power factor} = \frac{\text{true power}}{\text{apparent power}} = \frac{\text{watts}}{\text{volt-ampers}} = \frac{\text{kW}}{\text{kVA}}$$

For both the inductive and capacitive circuits, the apparent power is

$$\begin{aligned} P &= E \times I \\ &= 100 \times 50 = 5000 \text{ volt-amperes} \end{aligned}$$

The true power is zero since the net change from electrical energy in both of these types of circuits is zero. For such circuits, the power factor is

$$\text{PF} = \frac{\text{true power}}{\text{apparent power}} = \frac{0 \text{ Watts}}{5000 \text{ Volt-Amperes}} = 0$$

For the resistance circuit, however, the power factor is

$$\text{PF} = \frac{\text{true power}}{\text{apparent power}} = \frac{5000 \text{ Watts}}{5000 \text{ Volt-Amperes}} = 1$$

In the resistance circuit, all the energy delivered to it was converted into heat or mechanical energy and none of it was returned. In the inductive and capacitive circuits, all of the electrical energy delivered to them was returned to the source. The power factor of a circuit might therefore be defined as a factor which indicates how much of the electrical energy delivered to the circuit is converted by it into some other form of energy which is not returned.

Power in Combined Circuits

Figure 11–3 shows the curves for voltages, currents, and power in circuits having a combination of resistance, inductance, and capacitance. In Figure 11–3a, resistance, inductance, and capacitance of equal values are connected in parallel. From the curve, as also from Chapter 9, it is seen that the current flowing through the inductance and the current flowing through the capacitance are exactly equal and opposite and will therefore cancel each other. The true power in the circuit, as well as the apparent power, will therefore be the product of the voltage and the current flowing through the resistance. The power factor of this circuit will therefore be 1 (or unity or 100%).

Figure 11–3b shows a circuit having equal values of resistance and inductance in series. The currents flowing through each are the same, but the voltage drop across the resistance is displaced by a quarter-cycle from that across the inductance. The sum of these two voltage drops, which will be the voltage applied across the terminals of the conductor, is shown as the dashed line in the curve. If the true power is compared to the apparent power in this circuit, the power factor of the circuit will be found to be 0.707 or 70.7%.

IMPEDANCE

As mentioned previously, the total obstruction to the flow of electricity in an alternating-current circuit is called impedance. It represents the total effect of the obstructions to the flow of electricity presented by resistance, inductance, and capacitance, taking into account the fact that these several obstructions do not occur simultaneously.

MEASURING INSTRUMENTS

Modern measuring instruments for the determination of values of electrical current, voltage, power, and energy are all based on the interaction between a magnetic field and a conductor carrying a current. In their simplest forms, they are small but accurate motors. In measuring current, voltage, and power, the tendency for rotation of the small motors is pitted against a spring. In measuring energy, the motor revolves and the number of revolutions counted.

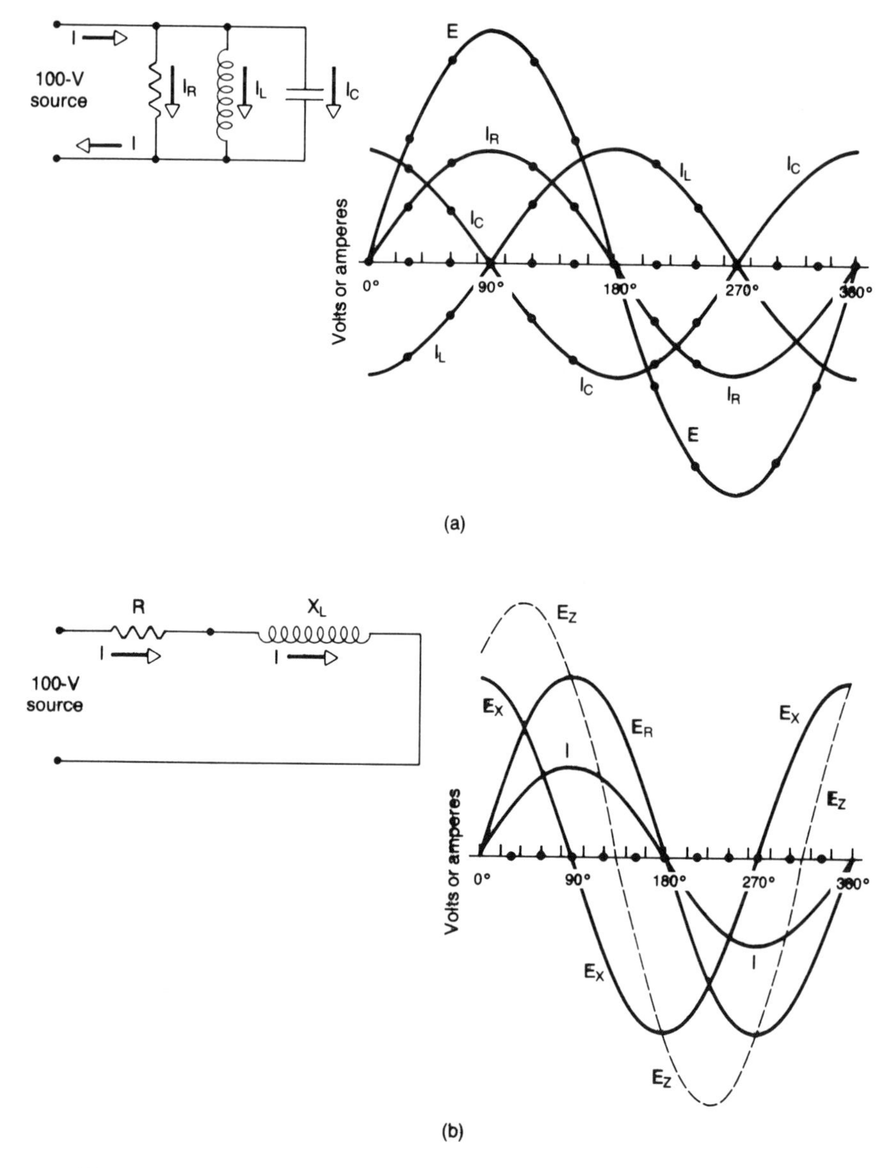

Figure 11–3 Cycle of voltage and current values: (a) resistance, inductance, and capacitance in parallel; (b) resistance and inductance in series.

Ammeter and Voltmeter

Figure 11–4a shows a cross section of an elementary ammeter or voltmeter. The magnetic field in this case is provided by a permanent magnet having certain predetermined characteristics. The current flowing through the coil produces a tendency for the coil to rotate. Springs on either side of the coil prevent it from rotating. However, the force tending to make the coil rotate will succeed in making it move. The amount the coil moves, as indicated by a pointer, is a measure of the current flowing.

A voltmeter is simply an ammeter having a fixed (and usually very large) resistance. By Ohm's law, voltage is equal to the product of current and resistance. If the resistance is constant, the value of a current is a measure of the applied voltage. This is exactly how a voltage is measured; the scale of the ammeter in this case is calibrated to read volts directly and the instrument is known as a voltmeter.

Wattmeter

In a wattmeter, in place of the permanent magnet to supply a magnetic field, an electromagnet is substituted. The electromagnet is connected in the circuit so that the magnetic field will be dependent on the current flowing through the circuit. The voltage across the movable coil will cause a current to flow through it which will cause the coil to have a tendency to rotate. This rotation will vary not only with the voltage across the coil, but also with the current flowing in the electromagnet. The rotating effect in this case is also pitted against a spring. In this case, however, the amount the coil moves, as indicated by the pointer, is a measure of the power flowing.

Watt-Hour Meter

In the watt-hour meter, for measuring electrical energy, a disk takes the place of the movable coil. An electromagnet having two coils is constructed as shown in Figure 11–4c, so as to allow the disk to rotate between a pair of poles. One coil is connected to measure current while the other is connected to measure voltage. The resultant magnetic field set up by both these coils induces eddy currents in the disk and acts on it in much the same way as if it were a conductor. The disk will therefore rotate, not having any springs to prevent it from doing so. The speed at which it rotates will depend on the power flowing. A train of gears and a set of dials are arranged to count the number of revolutions made by the disk. The number of revolutions is a measure of the electrical energy changed to some other form by the electrical circuit. All of these instruments are calibrated to measure effective values of current and voltage.

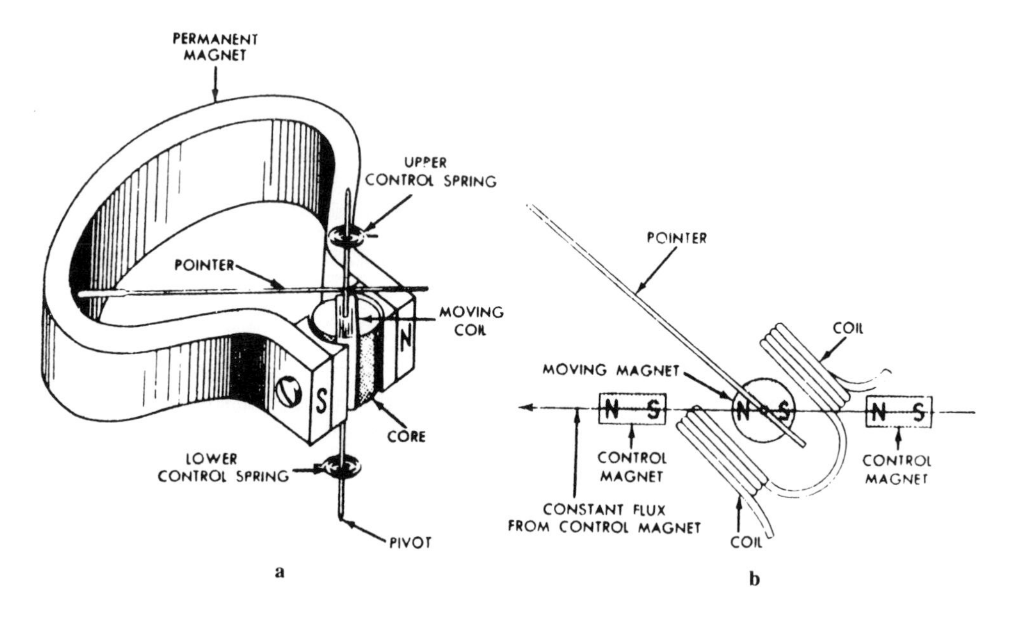
PERMANENT MAGNET
UPPER CONTROL SPRING
POINTER
MOVING COIL
N
S
CORE
LOWER CONTROL SPRING
PIVOT
a
POINTER
COIL
MOVING MAGNET
N S
N S
N S
CONTROL MAGNET
CONTROL MAGNET
CONSTANT FLUX FROM CONTROL MAGNET
COIL
b

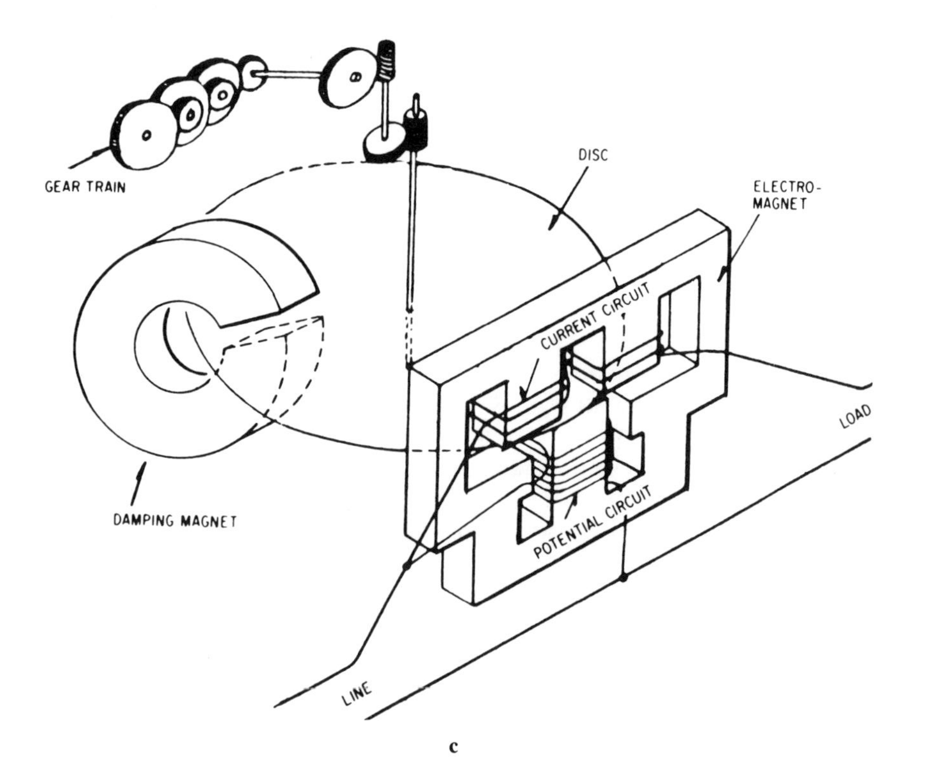
GEAR TRAIN
DISC
ELECTRO-MAGNET
CURRENT CIRCUIT
LOAD
POTENTIAL CIRCUIT
DAMPING MAGNET
LINE
c

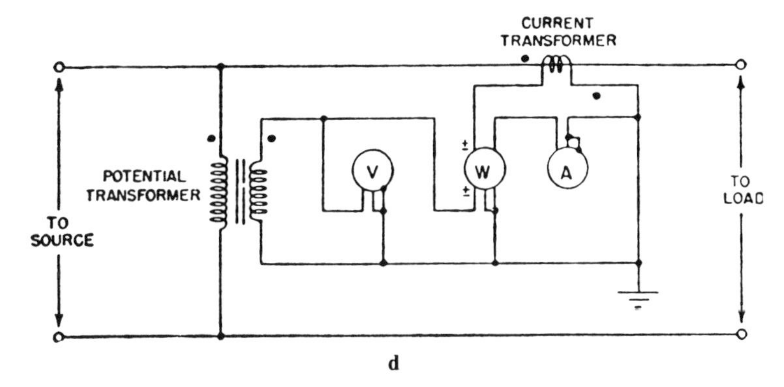

d

Figure 11–4 (a) Construction view and (b) schematic diagram of elementary ammeter or voltmeter. (c) Schematic diagram of watt-hour meter showing electromagnet. (d) Diagram showing method of connection for ammeter, voltmeter, and wattmeter in circuit.

Method of Connecting Instruments

Since an ammeter measures current, it must be connected in series with the circuit so that the current flowing in the circuit will also flow through it. A voltmeter must be connected across or in parallel with the circuit if it is to measure the electrical pressure applied to the circuit. Similarly, the "current" coil in a wattmeter or a watt-hour meter must be connected in series with the circuit, the "voltage" coil across or in parallel. A diagram showing how an ammeter, voltmeter, and wattmeter or watt-hour meter are connected in a circuit is shown in Figure 11–4d. This applies equally well to direct-current and alternating-current circuits.

INSTRUMENT TRANSFORMERS

When values of current or voltage are large, or when it is desired to insulate the meter from the circuit in which the electrical values are to be measured, an instrument transformer is used. In measuring current of high value, a *current transformer* (CT) is used. The ratio of transformation is such that the high circuit current, which in this case is the primary of the transformer, is reduced to a small current in the secondary connected to the ammeter. Current transformers have been practically standardized so that the secondary will produce 5 amperes, which is full scale for the ammeter associated with it.

Similarly, *potential transformers* (PT) have a fixed ratio of primary to secondary voltage and are usually built for operation at a secondary voltage of about 150 or 300 volts.

Instrument transformers differ from distribution or power transformers

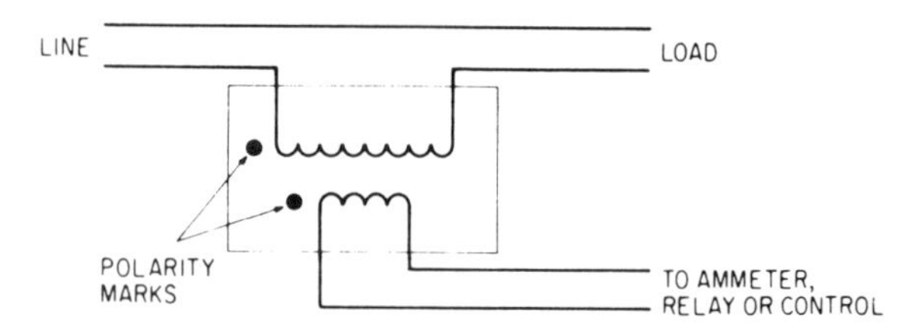

Figure 11–5 Current transformer and ammeter.

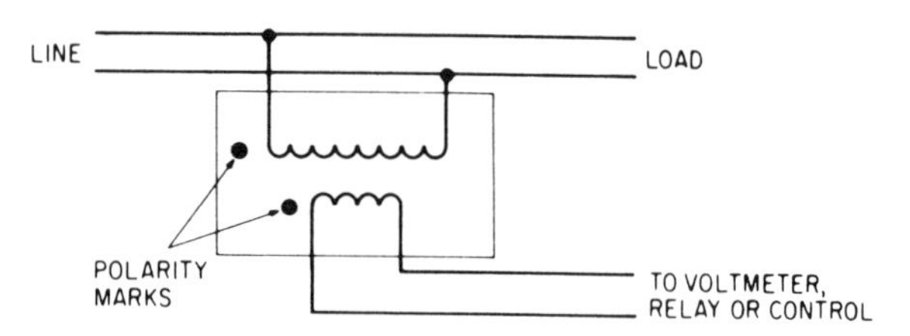

Figure 11–6 Potential transformer and voltmeter.

only in that they are of small capacity and are designed to maintain a higher degree of accuracy in their ratios of transformation than is necessary in the other types of transformers. Connections for both current transformers and potential transformers are shown in Figures 11–5 and 11–6.

REVIEW

- Because of the varying values of voltage and current in alternating-current systems, the effective value of these quantities will be less than the peak values; the effective value is 0.707 of the peak values.
- Comparable direct-current values will therefore be 0.707 of the peak alternating-current values. Hence the insulation of conductors carrying a direct current need be only 0.707 that of one carrying an alternating current.
- Power, or the rate of expending energy, is the product of voltage and current acting together. For direct-current systems, this is the product of the voltage of the circuit and the current flowing in it in amperes.
- In alternating-current systems, because of inductance and capacitance, the two values do not always act totally in conjunction. Only that portion of these acting together produce true power, expressed in watts. The product of the voltage and current is known as the apparent power, expressed in volt-amperes.
- The ratio between the true power and the apparent power is known as the power factor of the circuit.

- Measuring instruments are generally based on the interaction of the magnetic fields of two or more conductors. This interaction generally produces a tendency for one conductor to move in relation to the others, somewhat similar to motor action. A spring usually limits the motion of the conductor.
- An ammeter measures current in amperes; a voltmeter measures pressure in volts; a wattmeter measures power in watts; a watt-hour meter measures energy in watt-hours.
- The elements of these meters measuring current are connected in series in the circuit; those measuring voltage are connected in parallel.
- Where the values of the two basic quantities are rather high, instrument transformers are employed: current transformers to measure a fixed proportion of current flow and potential transformers to measure a fixed proportion of the voltage applied to the circuit.

STUDY QUESTIONS

1. Why cannot voltage and current in an alternating-current system be as readily evaluated as in the direct-current system?
2. What is meant by the effective values of voltage and current? How are they determined?
3. What is meant by "power" in an electrical circuit?
4. How does power vary throughout an alternating-current cycle in a resistance circuit; in an inductive circuit; in a capacitive circuit; in a combined circuit.
5. What is the "power factor" of a circuit? How is it determined?
6. What is the power factor of a circuit containing resistance only? Inductance only? Capacitance only? A combination of resistance, inductance, and capacitance?
7. What is the principle of operation of elementary measuring instruments?
8. What is an ammeter, voltmeter, and wattmeter, and how do they function? How are they connected in a circuit?
9. What is a watt-hour meter, and how does it function? How is it connected in a circuit?
10. What are instrument transformers, and when are they used? How are they connected in a circuit?

chapter 12

Vector Method of Representation

PRINCIPLES OF VECTOR METHOD

There are two methods that can be used for the addition of alternating voltages and currents of sine-wave form. One method consists of adding the instantaneous values throughout the cycle, as was done in Chapter 11 for circuits containing resistance and reactance. Such a method results in a correct addition but is rather cumbersome and slow. There is another method that gives equally correct results but is easier and quicker. This method is known as the *vector representation method* or, more simply, the *vector method*.

In Figure 12–1a, assume that the line *oa* represents a coil rotating in a uniform magnetic field, the assembly representing a simple electric generator. The voltage induced by the coil at different points along its path of rotation may be represented by the vertical lines drawn from the points marked 1, 2, 3, and so on, to the middle horizontal line. Thus, at position 0 the voltage induced is zero, while at position 3, it is a maximum. If the lengths of the vertical lines are laid off at regular intervals above and below the centerline, and the points are connected by a smooth curve, the result will be a wave which shows the instantaneous values of voltage induced in the coil as it rotates in the magnetic field. As mentioned previously, this wave is known as a *sine wave*.

Another such wave, drawn in a similar fashion but starting with the line

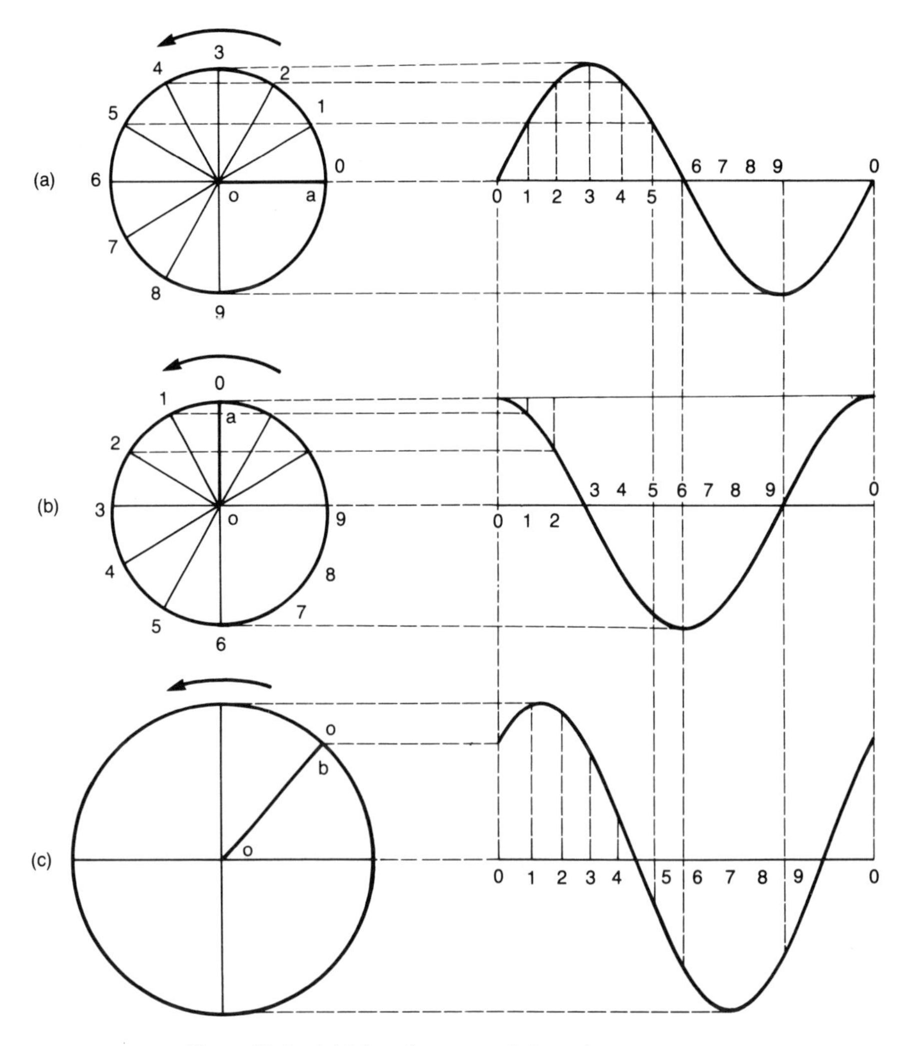

Figure 12–1 Addition of vectors and alternating-current waves.

oa in a vertical position, is shown in Figure 12–1b. Here the sine wave drawn is found to lag the first one by a quarter-cycle or 90 degrees. It is also to be noted, in Figure 12–1a and b, that the length of the line *oa* also represents the maximum value of the sine waves.

VECTORS

If both sine waves are superimposed and added, as shown in Figure 12–1c, it is found that the resultant wave is also a sine wave and can be produced by

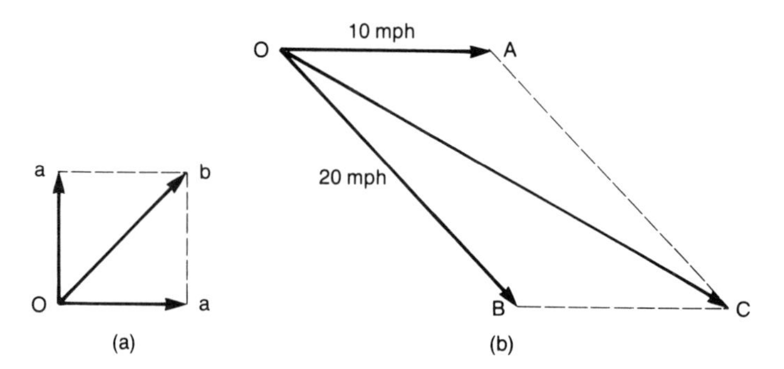

Figure 12–2 Addition of vectors.

revolving a line *ob* starting at the position shown, or 45° apart from the original lines *oa*. Instead of adding the instantaneous values of the sine waves, it is now only necessary to add the two lines *oa* at right angles to each other in the fashion shown in Figure 12–2a. The result will be the line *ob*, which will be correct not only in magnitude, but also in the position showing the relative displacement between each of the three waves. The lines *oa* and *ob* are referred to as *vectors*.

ADDITION OF VECTORS

This system may be compared to a floating object. Assume that it is drifting downstream at a speed of 10 miles per hour in the direction as indicated by *OA* in Figure 12–2b, and is blown by the wind at a speed of 20 miles per hour in the direction as indicated by *OB*. The resultant speed and direction of the object will then be *OC*. This can be found by replotting one of the lines (such as *OB*) at the end of the other (in position *AC*) and drawing the resultant *OC*. Lines *OA*, *OB*, and *OC* may also be called vectors.

Thus, by adding vectors in the manner shown, the resultant voltage or current may be readily found. In this method, a single straight line (called a vector) is used instead of the graph or curve of the instantaneous values to represent each voltage (or current). Each vector is drawn to a selected, convenient scale; that is, the length of the vector represents the amount of voltage (or current). The relative directions in which the vectors are drawn are determined by the angles of displacement, or phase difference, between the voltages or currents represented by the vectors. While the vectors shown represent the maximum instantaneous values, they may also represent the effective values since one value is a fixed proportion of the other.

Vector quantities may therefore be defined as quantities having both magnitude and direction. Quantities having magnitude only are known as scalar quantities. The conventional method of adding vectors is shown in Figure 12–3. It is desired to add two currents I_1 and I_2 lagging the common supply voltage

E. The resultant current I_3, both in magnitude and direction, can be found by completing the parallelogram $OACB$, which has the current I_1 as two of the parallel sides (OA and BC) and the current I_2 as the other pair of sides (OB and AC). The diagonal of this parallelogram is the resultant current (OC).

Component Method

A more accurate method is to resolve the current vectors, into vertical and horizontal components, and from the combined components determine the resultant. A little study of Figure 12–3 will show that OJ, the horizontal component of I_3, is equal to OG plus OD, and likewise the vertical component OK is equal to OF plus OH. Since the components may be found mathematically, more accurate results can be obtained than by the graphical method.

The actual current may therefore be considered as having two components, one due to the inductance of the line or equipment in the circuit and which lags 90 electrical degrees or a quarter-cycle behind the voltage. As explained in a previous session, the value of this lagging current is zero when the voltage has reached its maximum value. This lagging current is called the *reactive current*. The other component is known as the *active current*, and it is in phase with the voltage. This active current and the voltage reach maximum values simultaneously.

The actual line current may therefore be said to be the resultant of the reactive and active currents; furthermore, it is the current that would be measured by an ammeter in the circuit. Since there is one component lagging 90 electrical degrees or at right angles to the voltage; the resultant or actual line current, of which this component is a part, must consequently be out of phase with the voltage and lag behind it. The degree or amount that it lags depends upon the magnitude of this reactive current component and is a measure of the power factor.

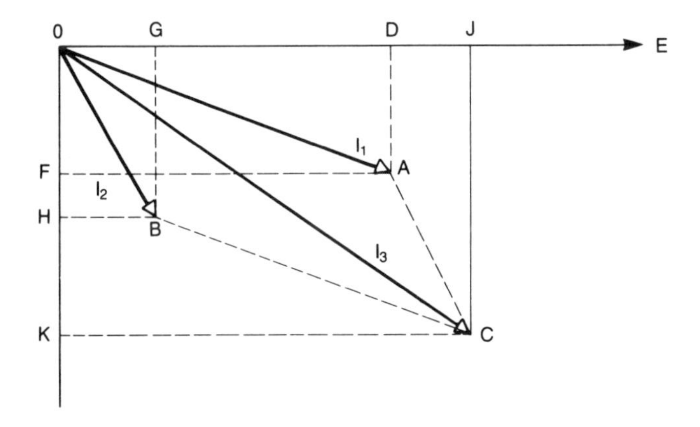

Figure 12–3 Component method of vector addition.

SUBTRACTION OF VECTORS

In subtracting vectors, reverse one and add it to the other. Figure 12–4 shows the same vector diagram as Figure 12–3, except that I_2 is subtracted from I_1, giving the resultant I_4.

POWER FACTOR

In Figure 12–5 the voltage is represented by the line OE. The line OI represents the actual line current as measured by an ammeter in the circuit, and lags behind the voltage by the angle ϕ. The active current component, in phase with the voltage, is OA and the reactive component is AI, 90° out of phase with the voltage.

If each of these three components of current is multiplied by the voltage, then:

1. Line current × voltage = apparent power (volt-amperes)
2. Active current × voltage = real power (watts)
3. Reactance current × voltage = reactance power, or watt-less power (measured in volt-amperes).

The real power is what is measured by a wattmeter (or watt-hour meter).

As pointed out before, the power factor is the ratio of the real power to

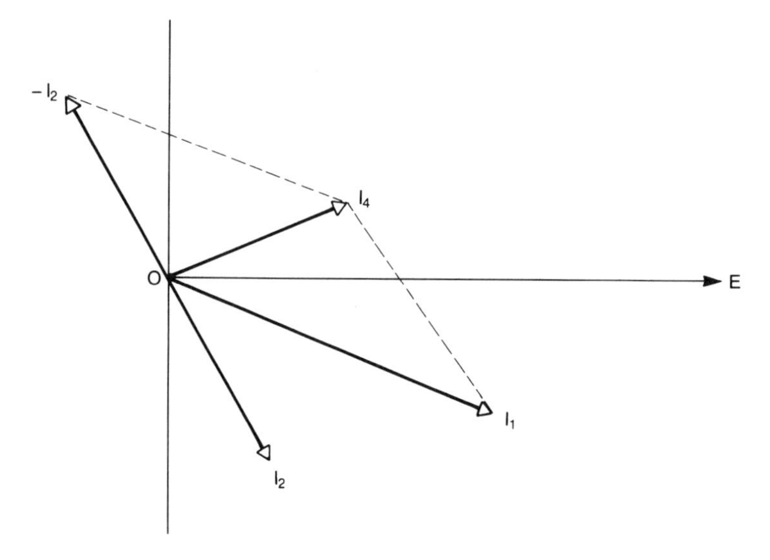

Figure 12–4 Subtraction of vectors.

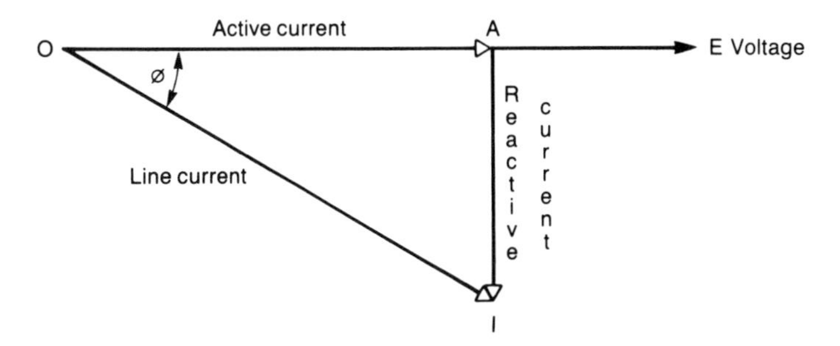

Figure 12–5 Power factor representation.

the apparent power. If both these quantities are divided by the voltage, the validity of the ratio remains unchanged. The power factor therefore may also be expressed as the ratio of the active current component to the actual line current. From Figure 12–5,

$$\text{power factor} = \frac{\text{active current}}{\text{line current}} = \frac{OA}{OI} = \text{cosine } \phi$$

The expression cosine ϕ is a mathematical expression that helps determine the parts of a right triangle; it expresses the relation between the side adjacent to the angle ϕ, OA, as compared to the hypotenuse of the triangle, OI.

REFERENCE VECTORS

In the preceding examples, the voltage was common to both currents and was plotted as the reference vector. The conventional rotation of vectors is counterclockwise. The direction of lag is clockwise and the direction of lead is counterclockwise. Any number of currents supplied by the same voltage can be added or subtracted on the same vector diagrams.

Circuits Having Current as Reference Vector

Consider again the current I, the voltages E_r and E_x of the circuit studied in Chapters 9 and 11 (Figure 12–6a), having a resistance and an inductance in series. Each of these three quantities can be represented by a vector as shown in Figure 12–6b. The current vector I is drawn to the right from the starting point O. Any convenient scale can be used and the arrowhead is used to indicate the ‘‘direction.’’ The voltage E_r is drawn in the same direction as the current since the two are in phase. As the voltage E_x is 90° ahead of I, the vector for E_x is drawn vertically upward, making an angle of 90° with the current vector. It must be drawn to the same scale as that selected to represent E_r.

The addition of the two voltages is easily made by adding the vectors.

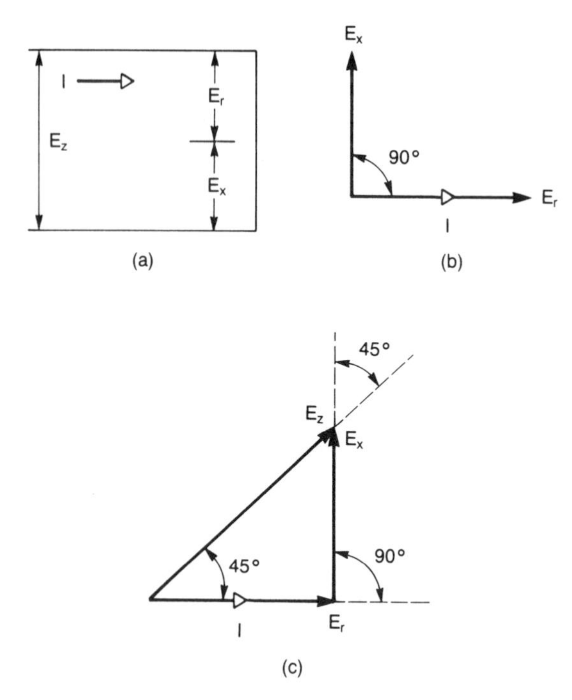

Figure 12–6 Vector diagram for resistance and inductance in series.

The addition is illustrated in Figure 12–6c. The vector for E_x is drawn vertically at the end of the line representing E_r, placing its tail end on the head end of the E_r vector and drawing it in the proper direction and of the proper length. The sum of the two, $E_r + E_x$, is represented by the distance from the starting point O to the end of the last vector added. This is the line E_z in Figure 12–6c. Its length can be determined by measurement. If E_r and E_x both represent 115 volts, then E_z represents a voltage of 162.6 volts.

In the vector diagram used above, a single straight line, the vector E_r, was drawn to scale to represent the 115-V sine wave, alternating voltage E_r of Figure 11–3b. Another single straight line, the vector E_x, was drawn to the same scale to represent the 115-volt, sine wave, alternating voltage E_x of Figure 11–3b. The sum of these two straight lines is another single straight line, the vector E_z, which represents, to the same scale, a sine-wave alternating voltage, which by measurement is found to be 162.6 volts. It should be remembered that each vector stands for a sine-wave alternating voltage (or current). The angles between the vectors represent time angles. These angles correspond to the angles of phase difference between the curves I, E_r, and E_z of Figure 11–3b.

In Figure 12–6c, the vector I is shown, although it is not necessary to perform the addition of E_r and E_x. Extensions (in dashed lines) are also shown on some of the vectors in order to show more clearly the angles of phase difference. Remembering that each vector represents a complete sine-wave alternating electrical quantity (a voltage or current), all of the information obtainable from Figure 11–3b can be obtained from the vector diagram of Figure 12–6c. I and E_r are in phase, E_x is 90° ahead of them and E_z is 45° ahead of them. The length of each vector determines (to scale) the effective value of the electrical quantity it represents.

In line-drop problems, the current is assumed to be common to all parts of the circuit and is used as the reference vector. In Figure 12–7 the load current I lags the voltage E by an angle θ at the receiving end of the feeder. There is a voltage drop IR across the resistance of the line which is in phase with the current, and a voltage drop IX across the reactance of the line which is ahead of the current 90 degrees. The IR drop is added to the receiving voltage in the position CD, and the IX drop is added vertically at point D as DF, giving the vector OF (or $E'L$) the required sending-end voltage. The actual voltage drop in the line is IZ (or line CF), where Z is the impedance of the line and is expressed in ohms. The voltage drop IZ is not always in phase with the voltage of the load and the net drop is the numerical difference between E' and E and is usually less than IZ.

In Figure 12–7 it will be seen that the current lags the sending-end voltage E' by a greater angle than it does the receiving voltage E, or, in other words, the power factor at the sending end is lower. Multiplying OA by the current I gives the watts supplied to the load. Adding the IR drop times I giving the power loss (I^2R loss) in the line, the total is the power supplied by the sending end.

In the case of leading power factor, the line drop is small and may even be negative (i.e., a voltage rise may occur). In Figure 12–8 the net line drop $E' - E$ is small and would reduce to zero if the angle of lead were increased.

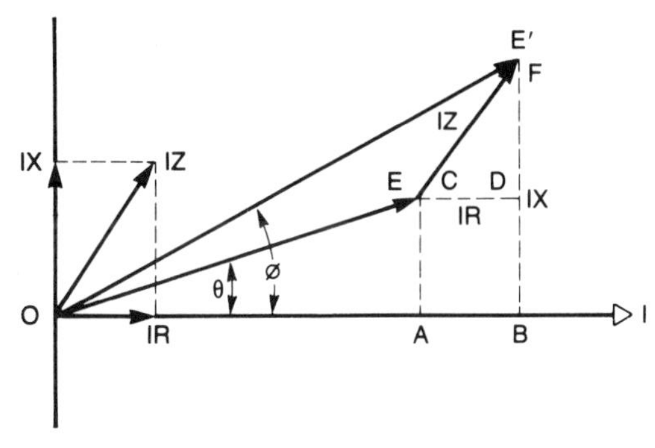

Figure 12–7 Vector diagram for line drop problem (I lags E).

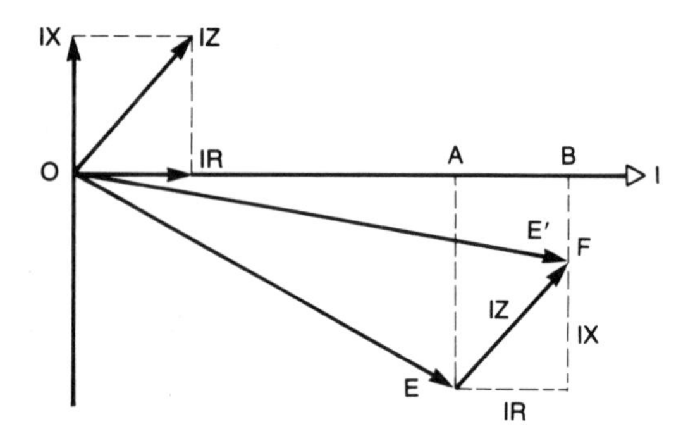

Figure 12–8 Vector diagram for line drop problem (I leads E).

Circuits Having Voltage as Reference Vector

The vector for the voltage E and the three currents I_r, I_x and I_c for the circuit shown in Figure 11–3a of resistance, inductance, and capacitance are illustrated in Figure 12–9. In this diagram L_R is in phase with E, I_C is 90° ahead

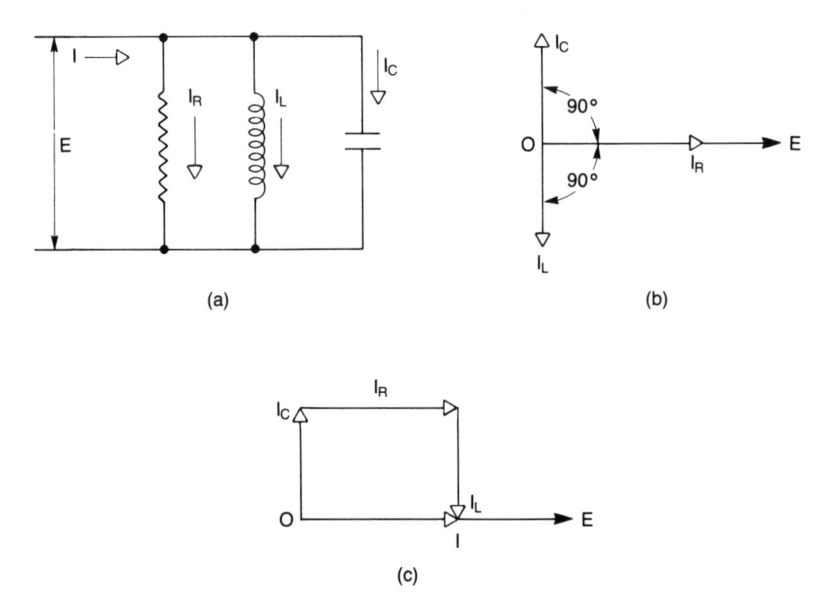

Figure 12–9 Vector diagram for resistance, inductance, and capacitance in parallel.

of E, and I_L is 90° behind E. The addition of the three currents $I_C + I_R + I_L$ is indicated in Figure 12–9c. Starting at 0 the vector I_C is drawn to any convenient scale in the proper direction (upward, 90° from E); the vector I_R is added on, being drawn to the same scale (to the right in the same direction as E with which it is in phase); the vector I_L is added on, being drawn to the same scale and in the proper direction (downward, 90° behind E). The sum of all three vectors is represented by the distance (and direction) from the starting point O to the tip of the last vector added. For this circuit this sum is seen to be the same in amount and direction as the vector for I_R alone. As in all vector diagrams used to represent electrical quantities, each current vector represents a sine-wave alternating current. The addition of the three sine-wave currents $I_C + I_R + I_L$ results in a sine-wave current represented in amount and direction (phase relationship) by the resulting vector I.

POLYPHASE SYSTEMS

There are two types of alternating-current circuits: (1) single-phase circuits and (2) polyphase (or multiphase) circuits. In single-phase circuits, only one phase or set of voltages of sine-wave form is applied to the circuits and only one phase of sine-wave current flows in the circuits. In the polyphase circuits, two or more phases or sets of sine-wave voltages are applied to the different portions of the circuits and a corresponding number of sine-wave currents flow in those portions of the circuits. The different portions of the polyphase circuits are usually called the *phase*. They are usually lettered to identify them, as the A phase, the B phase, and so on. The voltages applied to the separate phases of the circuits are correspondingly referred to as the A-phase voltage, the B-phase voltage, and so on. The phase currents of the different portions are also correspondingly identified as the A-phase current, the B-phase current, and so on.

The fundamental principles of the flow of alternating currents are the same whether applied to polyphase or single-phase circuits. The vector scheme of representation may also be applied to polyphase systems.

The voltages for polyphase systems are supplied from polyphase (multiphase) generators. Each phase of voltage is generated in a separate coil (or in coils connected in parallel), the separate coils being arranged for connection in different ways to form the polyphase system.

Two commonly used methods of connecting the coils of three-phase generators to supply a three-phase system are shown in Figure 12–10. One method, Figure 12–10a, employs the "delta" connection; the other, Figure 12–10b, the "star" (or Y) connection. The vector diagrams showing the voltage relationships for these connections are also included in Figure 12–10.

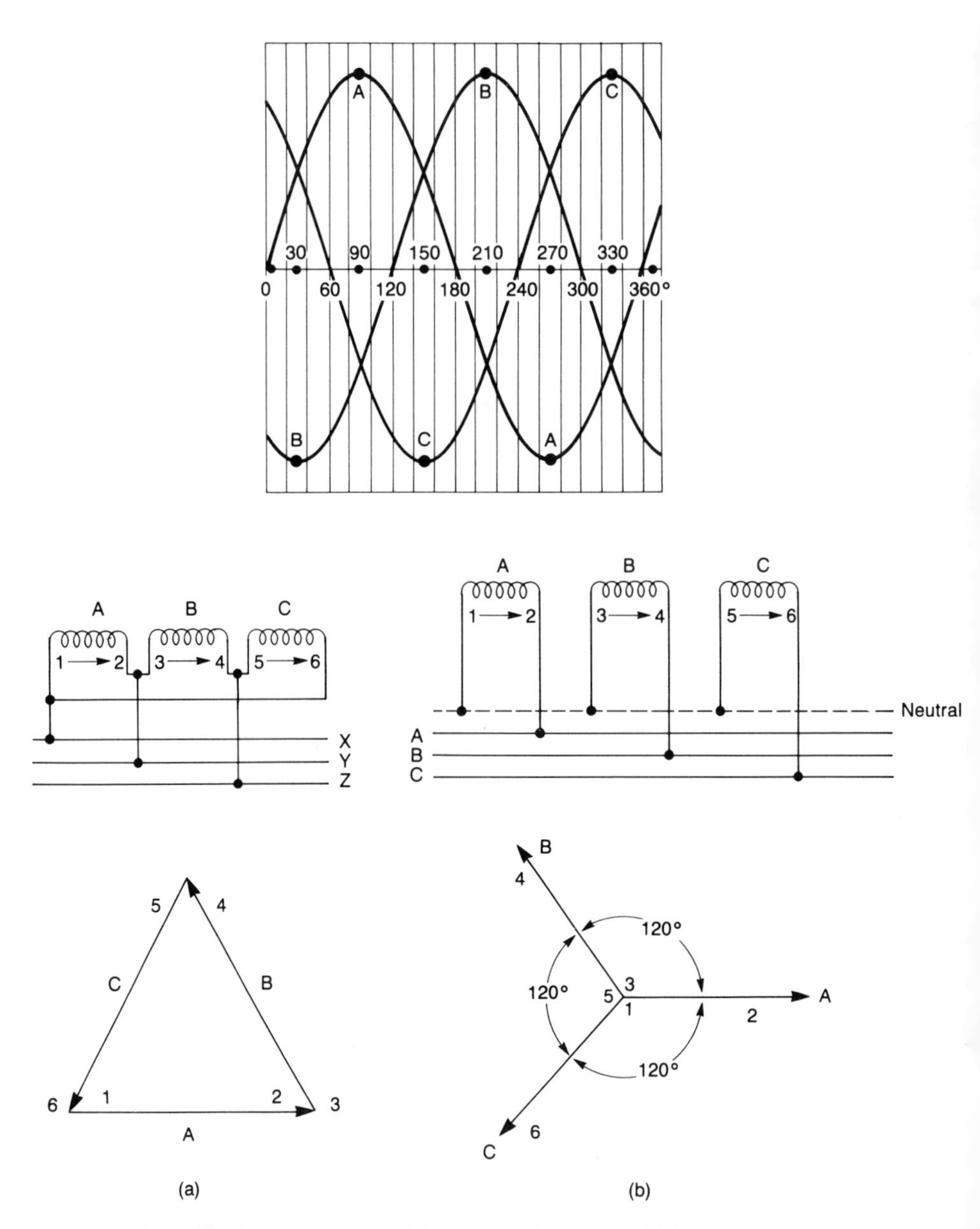

Figure 12–10 Vector diagrams: (a) three-phase delta system; (b) three-phase wye system.

REVIEW

- The vector method of handling alternating-current quantities constitutes a shorthand means of depicting their interrelationships.

- ✦ Rather than drawing sine-wave curves of appropriate magnitudes for voltage, current, and time relationships, these quantities are represented by straight lines.
- ✦ Their length represents their magnitude, and the angular displacement between the relative positions between the sine waves represents quantities.
- ✦ Quantities are added or subtracted by adding or subtracting the horizontal and vertical components of each of the vector quantities involved. The resultant is depicted by the net totals of the horizontal and vertical components.
- ✦ The power factor of an alternating-current circuit can be depicted by showing the voltage and current vector quantities displaced by an angle representing the leading or lagging of one quantity with reference to the other.
- ✦ Generally, series circuits will employ a vector representing current values as a reference vector for voltage quantities; multiple or parallel circuits will employ a vector representing voltage values as a reference vector for current values.
- ✦ Vectors are particularly useful in depicting relationships in polyphase circuits. These reflect not only interrelationships between voltage and current values for each phase, but such relationships between the several phases.
- ✦ The delta and wye connections of three-phase systems are based on their vector representations.

STUDY QUESTIONS

1. What is meant by a vector?
2. How are vectors added?
3. How are vectors subtracted?
4. What other method may be used to add or subtract vectors?
5. Into what components may line current be divided?
6. How may power factor be represented or determined in the vector method?
7. What is the conventional rotation of vectors? What is the direction of lag; of lead?
8. When is the current vector used as reference? When the voltage vector?
9. How may polyphase systems be represented by vectors?
10. Describe two commonly used methods of connecting the coils of three-phase generators and the vector diagrams showing the voltage relationships.

Appendix A

Effects of Harmonics on Motors

The voltage produced by an alternating current generator (or alternator) is generally taken to have a wave shape known as a sine wave. The frequency of such a voltage comprises the number of such waves, or cycle, per second – for the United States this is 60 cycles per second, the wave form of commercial generators, however, is never exactly a perfect sine wave (or sinusoidal) and under certain conditions may differ somewhat.

The voltage generated in the winding of the generator as it passes through the magnetic field of the poles of the generator will fluctuate as it passes from the field of one pole to that of the next pole. The voltage generated will begin to drop as it passes out of the field of the first pole and will rise again as it passes into the field of the next pole. The result is a distortion or "wriggle" in the shape of the alternating current wave produced. The distortion, although usually very small, about one or two percent, nonetheless creates a distortion form a pure sine wave. The distortion is in the form of other sine waves of different frequencies (and magnitudes) superimposed on the basic or fundamental sine wave; these are called harmonics of the voltage wave.

In alternating current circuits, the type of connected load also may have an effect on the voltage wave form. In one type of load, the voltage and current waves act in concert, or "in phase", and the current flowing is proportional to the voltage and the impedance (usually practically all resistance). This type load is referred to as a "linear" load (Figure A-1). In another type, the voltage and current do not act in concert and the current will vary disproportionally during each half cycles. These loads are referred to as "non-linear" loads and create the distortion in the wave form of the voltage and current waves, with the additional wave form created superimposed on the basic sine wave (figure A-2). The distortion produces waves of multiple frequencies within the basic (60 cycle per second) sine wave (Figure A-3).

Examples of linear loads include incandescent lamp, resistance type hearing, and constant speed induction and synchronous motors. Examples of non-linear loads are computers, battery chargers, electronic ballasts associated with gaseous type lighting devices, and apparatus and equipment in which the power supply is switched on and off as part of their operation.

Mathematically, the distorted sine wave may be broken down into a pure sine wave of the basic frequency and a series of other pure sine waves of different frequencies. Usually these are three times the basic, five times, seven times, etc. known as the third harmonic, fifth harmonic, seventh harmonic, etc. The magnitude of these harmonic voltages is small in comparison with that of the basic frequency, descending in magnitude as they vary from the basic frequency.

These harmonic voltages of different frequencies may be viewed as flowing in the same conductor. In a motor, these harmonic currents will produce their own alternating magnetic fields that will at different instances add to or subtract from that of the basic current. This will produce vibrations in the motor and the heat produced by them contribute to the (over) heating of the motor while not producing any mechanical power. Further, hysteresis losses are proportional to frequency and eddy current losses vary as the square of the frequency, both increasing because of the harmonic frequencies, producing additional losses in the core of the motor, in turn increasing the operating temperature of both the core and the surrounding windings.

The undesirable effects of heat and vibration caused by these harmonic voltages and currents may be mitigated by "draining" them off to ground before they reach the motor. This may be accomplished by inserting between the conductor and ground a "filter" consisting of an impedance of proper reactance and resistance of such values that the impedance will be low for circuits of harmonic frequencies but very high to the frequency of the main current (60 cycles per second).

While the discussion centers on a dingle conductor or single phase circuit, it is applicable to three-phase circuits. The filter impedance for each phase may be connected in delta between the three conductors with one terminal grounded or in wye with the common point connected to ground.

Another problem may be created by the harmonic currents. As non-linear loads (such as motors) often result in poor power factor performance, the resort has had to be corrective measures, generally with the installation of capacitors at the location of the load or nearby on the incoming supply circuit. These are usually connected between the phase conductor and the common neutral on a three-phase wye circuit. As capacitor reactance to the higher frequency harmonic voltages may result in comparatively large current flow (the impedance of capacitors decrease as the frequency increases), the size of the neutral conductor becomes important. Usual practice calls for a neutral conductor smaller in size that the phase conductors (on the basis that the three-phase currents, more or less balanced, tend to neutralize each other). The current flow from the high frequency voltages applied to the normal frequency capacitors may result in relatively high current flowing in the neutral as the phase currents may not neutralize each other but may add together. In such instances, the size of the neutral may be as large or larger than that of the phase conductors.

In addition to the problems with motor operation, harmonics may also cause fuses to blow on power factor correction capacitors, abnormal failure of fluorescent lighting ballast capacitor, and inefficient operation of adjustable frequency devices. When non-linear loads are added, harmonics produced by one load may affect other loads supplied by the same circuit. When harmonics are present, proper measure should be taken to prevent malfunction and damage to equipment.

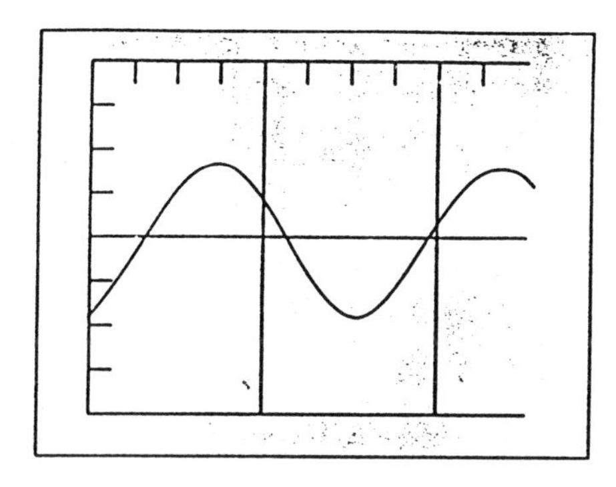

Figure A–1 Waveform in a linear load is a pure sine wave.

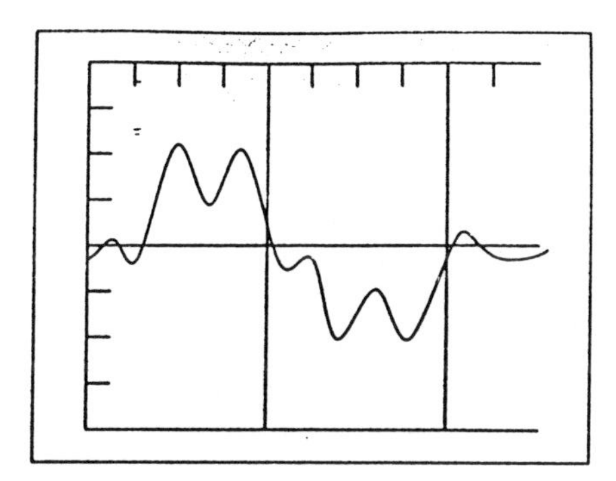

Figure A–2 Distorted wave form in a non-linear load, a combination of harmonic and basic sine waves.

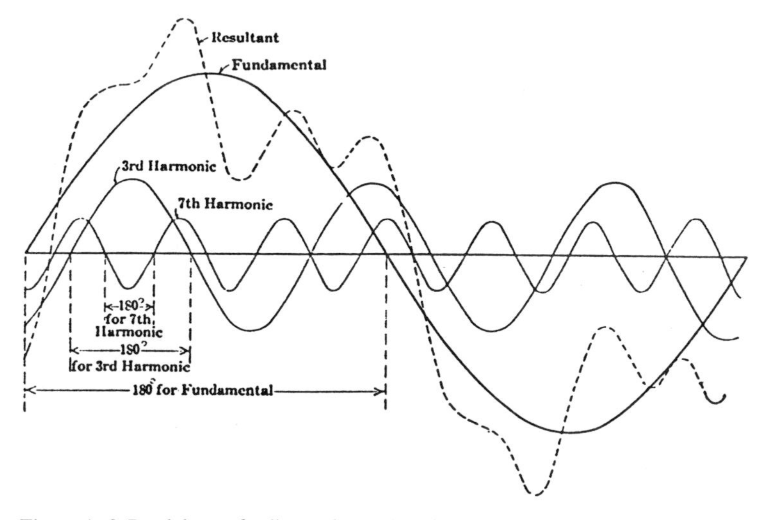

Figure A–3 Breakdown of a distorted wave into fundamental and harmonic sine waves.

Appendix B

Typical Specifications for Motor Purchase and Maintenance

Typical Specifications for Low-Voltage Motors: 460-V AC and 125-V DC

SCOPE

1.1. This specification covers the standard, general functional, and engineering requirements for electric motors of 460-V alternating-current and 125-V direct-current horizontal and vertical types.

1.2. The vendor is encouraged to communicate with the purchaser for an explanation of the intent of any portion of these specifications that the vendor feels are not clear.

STANDARDS

2.1. Except as otherwise specified, motors shall conform to the latest standards, including codes and technical definitions, whether or not specifically mentioned, of (in order of preference)

American National Standards Institute	ANSI
Institute of Electrical and Electronic Engineers	IEEE

American Society of Mechanical Engineers	ASME
National Board of Fire Underwriters—*National Electrical Code®*	NEC
National Electrical Manufacturers' Association	NEMA
Anti-Friction Bearing Manufacturers' Association	AFBMA

DESIGN

3.1. Design and construction of the motors shall be coordinated with driven equipment requirements, and motor vendor shall cooperate fully with manufacturers of driven equipment by furnishing information required for proper assembly and operation of the complete unit.

EXCEPTIONS

4.1. The vendor may suggest waiving certain requirements of this specification when, in his opinion, full compliance would not take advantage of particular design features or would involve unjustifiable cost or excessive delay in delivery. Suggestions should be accompanied by comparative performance and price data. No departures shall be made from this specification unless particular requirements are waived or modified in writing.

BEARINGS

5.1. Antifriction bearings for horizontal and vertical motors shall have a minimum rating life of 100,000 hours (10 years) for the bearing loading and speed conditions corresponding to the applications indicated.

5.2. Vendor may take exception to the specified bearing lift in the case of horizontal or vertical motors being furnished in NEMA rerated frame size, and state optimum bearing life available, for purchaser's approval.

5.3. Sleeve bearings when supplied for horizontal motors shall be ring-oiled type. Sleeve bearings shall be readily accessible and arranged so that the bearings may be inspected or replaced in a simple manner without disassembly of the motor or removal of the motor coupling. Split-bearing housings and split end bells shall be supplied as required to provide for above. Motor construction shall permit removal of the rotor without removing motor half-coupling.

5.4. When pressure oiling is required for horizontal motors, motor bearings shall be of the sleeve type arranged for pressure oiling supplied from lubrication system of the driven machine, with ring oiling for starting and emergency duty. The ring oiling shall be adequate for starting and continuous operation of the motor for at least ½ hour without pressure oiling system in operation.

5.5. All oil-lubricated bearings shall be provided with a visual means to field inspect the bearing oil rings for proper oil flow as well as an oil-level sight gauge which correctly indicates the level of oil within the bearing. The sight gauge shall be marked to show the oil level required for standstill and for running conditions.

5.6. Horizontal motors, except totally enclosed types, shall have means of measuring the motor air gap to check for sleeve bearing wear. This access shall not require disturbing the bearing housing.

5.7. Each horizontal motor with sleeve bearings shall be marked by die stencil or other permanent legend on the outside of the enclosure at full load and rated speed.

CONDUIT BOXES

6.1. Conduit boxes, accessibly mounted on the motor frame, shall be supplied for the motor main leads, resistance-temperature detector, and other thermal detector leads.

6.2. The conduit boxes for the motor main leads shall be capable of being turned through 360 degrees in 90 degree steps unless otherwise specified.

6.3. Openings from conduit box into motor shall be sealed on all totally enclosed motors.

6.4. All conduit boxes shall be weathertight. Gaskets of neoprene or approved equivalent shall be provided at cover joints and at joint between box and motor frame, unless these joints are otherwise weathertight. Conduit boxes shall be made corrosion resistant.

6.5. The following table specifies the minimum dimensions in inches for the motor main leads conduit boxes:

Horsepower	Length	Width	Depth
15 and below	5	4	5
20–50	8	6	6
60–150	11	9	10
200–250	14	11	14

DRIVE MEANS

7.1. The vendor shall furnish the motor complete with all accessories, such as half-coupling, pulley, sheave, or gear; this operation shall be performed with extreme care and without scoring or permanently deforming the shaft. The fit and method of installation shall permit removal of accessory without undue pressure and without scoring or deforming the shaft so that all ball, roller, or sleeve bearings may be removed and replaced without injury. The vendor shall adhere strictly to any dimensional requirements or other instructions relative to the installation of the accessory.

ENCLOSURE, VENTILATION, AND PROTECTION

8.1. Motors shall be self-ventilated unless otherwise specified.

8.2. When required, ventilating openings that are more than $\frac{3}{4}$ in. wide shall be protected with corrosion-resistant screens as follows: The openings in screens shall not exceed $\frac{1}{2}$ in.^{2} in area and shall be of such shape as not to permit the passage of a rod larger than $\frac{1}{2}$ in. in diameter, except that where the distance of exposed live or moving parts from the guard is more than 4 in., the openings may be $\frac{3}{4}$ in.^{2} in area and must be of such shape as not to permit the passage of a rod larger than $\frac{3}{4}$ in. in diameter. Screens are required to be removable when fixed screens would prevent convenient access for inspection and cleaning of motor windings and air gap.

8.3. All intake ventilating openings shall be furnished with screens or louvers.

8.4. All motors for outdoor service shall be furnished with screens or louvers on the outlet openings. Air outlet openings when located above the centerline of the motor shall be baffled.

8.5. All open, dripproof, and outdoor-type motors shall have screened drained openings to prevent accumulation of moisture inside. Totally enclosed motors shall be equipped with stainless steel or bronze drain plugs for the above purpose, and these openings shall be located to ensure drainage from all pockets of the motor casing.

8.6. Motors weighing more than 100 lb shall be provided with eye bolts, lugs, or other adequate means to insure safe lifting.

8.7. Horizontal motors shall be foot-mounted type, and vertical motors shall be base-mounted type, unless otherwise specified.

8.8. Motors shall be designed to permit convenient access for drilling vertically through motor feet or mounting flange for installation of purchaser's dowel pins after motors are mounted with driven equipment.

GROUNDING MEANS

9.1. Each motor shall have suitable means for attaching purchaser's grounding conductor in accordance with grounding requirements of the latest revision of the *National Electrical Code®*.

9.2. Two grounding terminals shall be located on the motor frame, on diagonally opposite sides of the motor; these shall be located on or near the feet of horizontal motors, and on or near the base of vertical motors. All terminals shall be visible.

9.3. When shielded or armored power cable is specified, a terminal shall be provided in the motor power lead box for grounding purposes.

INSULATION

10.1. All motors shall have a minimum of class B insulation.

LEADS AND TERMINALS

11.1. The main leads from the motor, resistance-temperature detector and thermal detector leads, and space heater leads where required, shall be brought into their respective conduit boxes; the leads or lead terminals shall be suitably marked and identified. Motor main leads or terminals shall be marked in accordance with ANSI Stanard for "Terminal Markings for Electrical Apparatus."

11.2. Compression-type terminal connectors shall be provided by the vendor for termination of all internal leads from the motor. On motors above 30 hp, connectors shall be right angles. Connectors for internal leads shall be crimped on motors above 30 hp.

ROTATION

12.1. Direction of rotation is defined as clockwise or counterclockwise, as observed when viewing the front end of a horizontal motor (i.e., when looking at end of motor opposite the drive or shaft extension) or when looking down at the top of vertical motor.

12.2. When feasible, motors shall be designed for operation in either direction of rotation without physical change in the motor.

12.3. All motors shall have the specified direction of rotation marked by an arrow on the stator frame and stamped on the nameplate.

12.4. The corresponding terminal phase sequence shall be marked on the motor nameplate. If the direction of rotation is not specified, the direction of rotation corresponding to the phase sequence T_1, T_2, T_3 shall be marked on the motor.

STARTING DUTY

13.1. Squirrel-cage induction and synchronous motors shall be designed for full-voltage starting. They shall withstand all stresses and give satisfactory performance when started with their driven equipment connected, with supply frequency at rated value and with motor terminal voltage not more than 10% above or below rated voltage.

13.2. It is to be assumed that no starting compensators or other means for reducing starting current will be used by the purchaser.

TEMPERATURE DETECTORS

14.1. Motor windings shall be equipped with a minimum of two temperature devices embedded in the windings.

14.2. Bearings shall be equipped with tip- and head-sensitive resistance-temperature detectors, or with chromel-constantan thermo-couples, insulated as required by application.

14.3. Where bearings are monitored by a temperature detector, the vendor shall state the recommended alarm and trip temperature settings in degrees centigrade.

TEMPERATURE REQUIREMENTS

15.1. Motors shall be rated for continuous operation in an ambient temperature not exceeding 40°C at an altitude not exceeding 1000 meters above mean sea level.

15.2. Values of temperature rise offered by the vendor shall be winding temperature rise, as measured by the resistance method.

15.3. Temperature rise shall not exceed the NEMA (MG-1) standard temperature rise for the rating, enclosure, and insulation called for in the purchaser's specifications.

TORQUE-CURRENT REQUIREMENTS

16.1. Motors operating with rated terminal voltage and rated frequency shall have torque values in accordance with the applicable ANSI Standards, except that the vendor may propose a more efficient or less expensive design with lower values of torque possible.

16.2. Locked-rotor current shall have the lowest value consistent with good performance and economical design for the torque-current class, and preferably shall not exceed 650% of full-load current.

16.3. Except as noted, motors shall be of the type described as having normal starting torque with low starting current.

NAMEPLATES

17.1. For squirrel-cage induction motors, the nameplate shall include the following minimum information:

a. Type designation
b. Frame number
c. Horsepower output rating
d. Time rating
e. Temperature rise in degrees C at rated load
f. Rated load speed (RPM)
g. Voltage rating
h. Amperes at rated load
i. Service factor (marked "for operation at 40°C ambient")
j. Frequency rating
k. Number of phases
l. Direction of rotation related to terminal connections (if limited to one direction only, so state)
m. Antifriction bearings renewal part number
n. Code letter

17.2. For synchronous motors, the nameplate shall include the following minimum information:

a–m. Same as for induction motors List above.
n. Rated field amperes
o. Rated exciter voltage
p. Rated power factor

17.3. For direct-current motors, the nameplate shall include the following minimum information:

a–h. Same as for induction motors List above.
i. Winding type (shunt, series, or compound)
j. Antifriction bearings renewal part number
k. Shunt field current

DRAWINGS AND DATA

18.1. Vendor shall submit to purchaser outline drawings of motors. Outline drawings shall show overall dimensions and other essential dimensions required for coordination with driven equipment, foundations, conduits, dimensional location of shaft mounting holes, and conduit boxes. Drawings shall identify particular motors and corresponding driven equipment to which they apply.

18.2. Vendor shall furnish, for each item, complete nomenclature and characteristic data listed below. These data shall be included on the outline drawings or on attached sheets. General information required:

1. Manufacturer's type designation
2. Manufacturer's frame number
3. Code letter
4. Design letter
5. Rated output in horsepower
6. Time rating
7. Temperature rise in degrees Celsius at rated load
8. Synchronous speed in rpm
9. Full-load speed in rpm
10. Rated frequency in cycles per second
11. Number of phases
12. Rated voltage
13. Full-load current in amperes
14. Service factor (where applicable)
15. Winding type (squirrel-cage, etc.)
16. Insulation class
17. Enclosure and ventilation type, provisions for required screens, and whether motor is for indoor or outdoor service
18. Antifriction bearing style and manufacturer's catalog number
19. Rating life of antifriction-type bearings for actual bearing speed and loading conditions indicated
20. Conduit opening (electrical trade size) for purchaser's conduit connection
21. Motor internal lead sizes, terminal connector types, and sizes for internal leads and purchaser's cables
22. Direction of rotation, or if either direction permissible
23. Construction, dimensions, and location of grounding means
24. Approximate weight of separate part of motor weighing more than 100 pounds (rotor or stator)
25. Net weight of complete motor
26. Efficiency at rated voltage and frequency for 25, 50, 75, and 100% of rated load
27. Power factor at rated voltage and frequency for 25, 50, 75, and 100% of rated load
28. Locked-rotor current in amperes

18.3. Submission of drawings, operating and maintenance instruction books, test data, and spare parts lists shall be furnished to purchaser as part of specifications.

INSTRUCTION BOOKS

19.1. Vendor shall furnish a composite instruction book specifically covering installation, operation, maintenance, and complete field overhaul of motors. Installation instructions shall include procedure for checking alignment of motor shaft, coupling, and base. Instructions shall include lubricating details, recommended lubrication inspection and replacement periods, quantity of lubricant, and specifications for replacement lubricant.

FACTORY TESTS

20.1. Motor shall be given factory routine tests or Manufacturer's "Standard Commercial Test" to ensure that it is free from electrical and mechanical defects.

PROTECTION DURING SHIPMENT

21.1. All equipment shall be adequately protected during shipment to prevent damage, corrosion, or entrance of foreign matter.

Typical Specifications for Small Induction Motors, 30 hp and Below

SCOPE

1.1. This specification covers the standard, general functional, and engineering requirements for small, polyphase squirrel-cage induction motors.

1.2. The vendor is encouraged to communicate with the purchaser for an explanation of the intent of any portion of these specifications that the vendor feels are not clear.

STANDARDS

2.1. Except as otherwise specified, motors shall conform to the latest standards, including codes and technical definitions, whether or not specifically mentioned of (in order of preference)

American National Standards Institute	ANSI
Institute of Electrical and Electronic Engineers	IEEE
American Society of Mechanical Engineers	ASME
National Board of Fire Underwriters—*National Electrical Code®*	NEC
National Electrical Manufacturers' Association	NEMA
Anti-Friction Bearing Manufacturers' Association	AFBMA

DESIGN

3.1. Design and construction of motors shall be coordinated with driven equipment requirements, and motor vendor shall cooperate fully with manufacturers of driven equipment by furnishing information required for proper assembly and operation of the complete unit.

EXCEPTIONS

4.1. The vendor may suggest waiving certain requirements of this specification when, in his opinion, full compliance would not take advantage of particular design features or would involve unjustifiable cost or excessive delay in delivery. Suggestions should be accompanied by comparative performance and price data. No departures shall be made from this specification unless particular requirements are waived or modified in writing.

BEARINGS

5.1. Antifriction bearings for horizontal and vertical motors shall have a minimum rating life of 100,000 hours (10 years) for the bearing loading and speed conditions corresponding to the applications indicated.

5.2. Vendor may take exception to the specified bearing life in the case of horizontal or vertical motors being furnished in NEMA rerated frame size, and state optimum bearing life available, for purchaser's approval.

CONDUIT BOXES

6.1. Conduit boxes, accessibly mounted on the motor frame, shall be supplied for the motor main leads, resistance-temperature detector, and other thermal detector leads.

6.2. For motors rated one horsepower or more, the conduit boxes (for the motor main leads) shall be capable of being turned through 360 degrees in 90 degree steps unless otherwise specified.

6.3. Openings from conduit box into motor shall be sealed on all totally enclosed motors.

6.4. Conduit boxes on motors for outdoor service shall be weathertight. Gaskets of neoprene or approved equivalent shall be provided at cover joints and at joints between box and motor frame unless these joints are otherwise weathertight.

6.5. Unless otherwise specified, the conduit boxes of NEMA type 1 weather-protected and totally enclosed motors shall have hubs or tapped openings provided for entrance of conduits of the size specified for each item.

6.6. When not otherwise indicated, conduit boxes for horizontal motors shall be in the standard location, that is, on the right-hand side when viewing the front end of the motor (i.e., when looking at end of motor opposite the drive or shaft extension).

ENCLOSURE, VENTILATION, AND PROTECTION

7.1. Motors shall be self-ventilated unless otherwise specified.

7.2. Air outlet openings when located above the centerline of the motor shall be baffled.

7.3. Screens for intake ventilating openings shall be furnished (except on motors for which air filters are specified and on totally enclosed motors).

7.4. The term "weather-protected" as used in this specification applies to motors subjected to outdoor conditioning.

7.5. All motors for outdoor installation shall be equipped with screens on the outlet openings.

7.6. The term "totally enclosed types" used in this specification includes totally enclosed nonventilated and totally enclosed fan-cooled types.

GROUNDING MEANS

8.1. Each motor shall have suitable means for attaching purchaser's grounding conductor in accordance with grounding requirements of the latest revision of the *National Electrical Code®*.

9.2. Two grounding terminals shall be located on the motor frame, on diagonally opposite sides of the motor; these shall be located on or near the feet of horizontal motors, and on or near the base of vertical motors. All terminals shall be visible.

INSULATION

10.1. All motors shall have class B insulation; all motors, unless totally enclosed, shall have encapsulated end turns.

LEADS AND TERMINALS

11.1. The main leads from the motor, resistance-temperature detector and thermal detector leads, and space heater leads where required, shall be brought into their respective conduit boxes; the leads or lead terminals shall be suitably marked and identified. Motor main leads or terminals shall be marked in accordance with ANSI Standard for "Terminal Markings for Electrical Apparatus."

11.2. Compression-type terminal connectors shall be provided by the vendor for termination of all internal leads from the motor. On motors above 15 hp, connectors shall be right angles; on motors below 15 hp, connectors shall be straight. Connectors for internal leads shall not be crimped for motors above 15 hp.

LIFTING MEANS

12.1. Motors weighing more than 100 pounds shall be provided with eye bolts, lugs, or other adequate means to ensure safe lifting.

MOUNTING

13.1. Horizontal motors shall be foot-mounted type, and vertical motors shall be base-mounted type, unless otherwise specified.

NAMEPLATES

14.1. Squirrel-cage induction motors shall have a nameplate including the following minimum amount of information:

a. Type designation

b. Frame number
c. Code letter
d. Design letter
e. Horsepower output rating
f. Time rating
g. Temperature rise in degrees C at rated load
h. Rated load speed (rpm)
i. Voltage rating
j. Amperes at rated load
k. Service factor (marked: for operation at 40°C ambient)
l. Frequency rating
m. Number of phases
n. Total weight of motor

ROTATION

15.1. Direction of rotation is defined as clockwise or counterclockwise as observed when viewing the front end of a horizontal motor (i.e., when looking at end of motor opposite the drive or shaft extension) or when looking down at the top of vertical motor.

15.2. When feasible, motors shall be designed for operation in either direction of rotation without physical change in the motor.

15.3. All motors shall have the specified direction of rotation marked by an arrow on the stator frame and stamped on the nameplate.

15.4. The corresponding terminal phase sequence shall be marked on the motor, preferably on the nameplate and inside the terminal conduit box. If the direction of rotation is not specified, the direction of rotation corresponding to the phase sequence T_1, T_2, T_3 shall be marked on the motor.

SERVICE FACTOR

16.1. Service factor, when required, shall correspond to permissible continuous overload operation at rated voltage and frequency with ambient temperature not exceeding 40°C and altitude not exceeding 1000 meters.

SOUND LEVEL

17.1. The vendor shall submit with his proposal anticipated (not guaranteed) noise levels for motors of similar ratings to those specified.

17.2. Sound-level values shall be given for the following octave bands:

Octave Band No.	Frequency Range in Cycles per Second
1	20–75
2	75–150
3	150–300
4	300–600
5	600–1200
6	1200–2400
7	2400–4800
8	4800–10,000

STARTING DUTY

18.1. Squirrel-cage induction motors shall be designed for full voltage starting. They shall withstand all stresses and give satisfactory performance when started with their driven equipment connected, with supply frequency at rated value, and with motor terminal voltage not more than 10% above or below rated voltage.

18.2. It is to be assumed that no starting compensators or other means for reducing starting current will be used by the purchaser.

TEMPERATURE DETECTORS

19.1. When specified, motors shall be furnished with thermocouple or thermostat protection for motor windings. Contacts shall be normally closed, and open at the high-temperature setting.

TEMPERATURE REQUIREMENTS

20.1. Motors shall be rated for continuous operation in an ambient temperature not exceeding 40°C at an altitude not exceeding 1000 meters above mean sea level.

20.2. Values of temperature rise offered by the vendor shall be winding temperature rise, as measured by thermocouple. If the motor is not equipped with this device, the resistance method may be used.

20.3. Temperature rise shall not exceed the NEMA (MG-1) standard temperature rise for the rating enclosure and insulation called for in the purchaser's specifications.

TORQUE-CURRENT REQUIREMENTS

21.1. Squirrel-cage induction motors operating with rated terminal voltage and rated frequency shall have torque values in accordance with the applicable ANSI Standards, except that the vendor may propose a more efficient or less expensive design with lower values of torque when possible.

21.2. Locked-rotor current shall have the lowest value consistent with good performance and economical design for the torque-current class, and preferably shall not exceed 650% of full-load current.

21.3. Except as noted, motors shall be of the type described as having normal starting torque with low starting current.

DRAWINGS AND DATA

22.1. Vendor shall submit to purchaser outline drawings of motors. Outline drawings shall show overall dimensions and other essential dimensions as required for coordination with driven equipment, foundations, conduits, dimensional location of shaft mounting holes, and conduit boxes. Drawings shall identify the particular motors and corresponding driven equipment to which they apply.

22.2. Vendor shall furnish for each item complete nomenclature and characteristic data as listed below. These data shall be included on the outline drawings or on attached sheets. General information required:

1. Manufacturer's type designation
2. Manufacturer's frame number
3. Code letter
4. Rated output in horsepower
5. Time rating
6. Temperature rise in degrees Celsius at rated load
7. Synchronous speed in rpm
8. Full-load speed in rpm
9. Rated frequency in cycles per second
10. Number of phases
11. Rated voltage
12. Full-load current in amperes
13. Service factor (where applicable)
14. Winding type (squirrel-cage, etc.)
15. Insulation class
16. Enclosure and ventilation type, provisions for required screens, and whether inside or outside motor service
17. Antifriction bearing style and bearing manufacturer's catalog number
18. Rating life of antifriction type bearings for the actual bearing speed and loading conditions indicated

19. Conduit opening (electrical trade size) for purchaser's conduit connection
20. Motor internal lead sizes; also, terminal connector types and sizes for internal leads and purchaser's cables
21. Direction of rotation, or if either direction permissible
22. Construction, dimensions, and location of grounding means
23. Approximate weight of any separate part of motor (rotor or stator) weighing more than 100 pounds
24. Net weight of complete motor
25. Efficiency at rated voltage and frequency at 50, 75, and 100% of rated load
26. Power factor at rated voltage and frequency for 50, 75, and 100% of rated load
27. Locked-rotor current in amperes
28. Maximum noise level in decibels for all motors

22.3. Submission of drawings, operating and maintenance instruction books, test data, and spare parts lists shall be furnished to purchaser as part of specifications.

INSTRUCTION BOOKS

23.1. Vendor shall furnish a composite instruction book specifically covering installation, operation, maintenance, and complete field overhaul of motors. Installation instructions shall include procedure for checking alignment of motor shaft, coupling, and base. Instructions shall cover lubricating details, including recommended lubrication inspection and replacement periods, recommended quantity of lubricant, and specifications for replacement lubricant.

FACTORY TESTS

24.1. Each motor shall be given factory routine tests or manufacturer's "Standard Commercial Test," as necessary to determine that it is free from electrical and mechanical defects.

PROTECTION DURING SHIPMENT

25.1. Equipment shall be adequately protected during shipment to prevent damage, corrosion, or entrance of foreign matter.

Motor Data

1 Location	Spec. No.
2 Furnished by	Date
3 Mark or item No.	Work order
4 Purchaser's requirements	Purchaser's requirements
5 Service	
6 Type	Bearing type
7 Number of units	Bearing temp. relay
8 Mounting	Bearing temp. device
9 Elec. characteristics V Ph Hz	Bearing service water °F psi
10 Synch speed rpm	Bearing service hours
11 Horsepower	
12 Service factor	
13 Enclosure	CT ratio
14 Insulation class	Surge capacitors
15 Insulation treatment	Minimum starting voltage %
16 Ambient temp. °C	Neut. leads avail. in term. box
17 Stator temp. rise °C	
18 Project elev. ft	Thermal limit curve
19 Location (indoor) (outdoor)	Speed–torque curve
20 Half-coupl. or sheave mtd by	Speed–current curve
21 Rotation**	Tests and inspection requirements
22 WK^2 of driven equip.	
23 Brkwy. torq. drvn. equip.	Motor assembly to be inspected
24 Oversize cond. box	Complete eng'g test
25 Cond. box location*	Standard commercial test
26 Space heaters, voltage, phase	Complete eng'g test witnessed
27 Type of cable and size	Max. overall sound press. level
28 Terminal lugs, type	unweighted dB
29 Stator high-temp. device	Sound test A-weighted
30 Adjustable slide rails	Octave band test
31 Soleplates	Third octave band test
32 Shaft (hollow, solid)	Sound test witnessed

Motor Data (*Continued*)

33 Coupling (self-release,	Sound level guaranteed
34 solid, nonreversing,	
35 adjustable, flexible)	
36 Vert. max. downthrust	
37 Vert. max. upthrust	
38 Vert. min. upthrust	
39 Vert. min. downthrust	
40 (with motor running)	
41 Side thrust	
42 Max. reverse speed	
43 Drain plug and vent	
44 Air intake	
45 Air exhaust	
46 Split end bells	
47 High-efficiency design required (yes) (no)	
48	
49	
50 Remarks:	
51 All performance data based on normal	
52 rated voltage and frequency	
53 Items 34–44 apply to vert. motors only	
54 Refer to either spec. E-290 or E-295	
55 for general requirements	
56 Refer to spec. E-982 for sound reqmts.	
57 * Mfgr. to supply data.	By: Date:

58 ** Viewed from end opposite coupling end.

Typical Specifications for Performance of Preventive Maintenance Service on Low- and Medium-Voltage Motors

DEFINITIONS

1.1. *Bidder*: a party submitting a proposal to fulfill the requirements of this specification

Vendor: the party accepting the overall responsibility for fulfilling the requirements of this specification

Purchaser: the party identified in the proposal to bid

Location: location where service is to be performed and identified in the proposal to bid

SCOPE OF WORK

2.1. This specification states the conditions and requirements for furnishing parts and services for the performance of preventive maintenance service on 125, 480, 2300, and 4160-V alternating-circuit electric motors. Services provided by the vendor shall consist of, but are not limited to, the following:

1. Steam cleaning
2. Dry out and baking
3. Varnishing
4. Machining
5. Mechanical and electrical checks and tests

EXCEPTIONS

3.1. If the bidder takes any exception to this specification:

1. Each exception shall be itemized and explained.
2. All exceptions shall be in a portion of the proposal headed "Exceptions to the Specification."

CODES AND STANDARDS

4.1. *IEEE 112-1984*: IEEE Standard Test Procedure for Polyphase Induction Motors and Generators

4.2. *IEEE 113-1985*: IEEE Guide: Test Procedure for Direct Current Machines

4.3. *IEEE 114-1982*: IEEE Standard Test Procedure for Single Phase Induction Motors

4.4. *IEEE 115-1983*: IEEE Guide: Test Procedures for Synchronous Machines

4.5. *IEEE C50.10-1977*: American National Standard General Requirements for Synchronous Machines

4.6. *NEMA (MG-1)1972*: National Electrical Manufacturers' Association Standards for Motors and Generators

GENERAL REQUIREMENTS

5.1. The repair shop of the vendor is expected to coordinate efforts with the purchaser. During periodic overhauls, turnabout time is critical to ensure equipment availability.

5.2. The repair shop of the vendor shall provide transportation from the purchaser's location to and from the repair shop. Where required, motor shafts should be blocked for shipment.

5.3. Purchaser is to be notified 24 hours prior to pickup and delivery.

5.4. Vendor's repair shop shall be responsible for diligently resolving any field problem associated with the subject motor serviced after its return and installation.

5.5. Notification to the purchaser is required when any standard condition is encountered outside the scope of this specification.

5.6. An invoice that includes a breakdown of time, material, and transportation charges for any work beyond the scope of preventive maintenance shall be sent to the purchaser.

TECHNICAL REQUIREMENTS: PREVENTIVE MAINTENANCE

6.1. Prior to disassembly and cleaning, perform visual inspection and incoming electrical tests: megger, phase continuity and phase balance, and polarization index tests. Measure air gap and record measurement. Check operation of any bearing thermocouples or winding thermal-indicating devices.

6.2. Disassemble motor, clean and inspect all components, including stator iron, end turn bracing, and rotor bars. Notify purchaser of any substandard condition and required repairs prior to initiating same.

6.3. Inspect bearings and journals. Record bearing and journal measurements. Clean lubricating oil system and refill with proper lubricant. When it is not advisable to refill lubricant, secure any information tag to each bearing, noting that it does not have lubricant.

6.4. Inspect power leads, strip heaters, if provided, and instrumentation wiring. All leads are to be verified and legibly tagged and identified.

6.5. Steam clean stator windings, varnish dip, and bake. Varnish insulation on all motors shall be equal to or better than provided, but shall be no lower than class B.

6.6. Clean air passages and replace missing filters.

6.7. Lathe check rotor for acceptable run-out.

6.8. Check that the mechanical center agrees with the electrical (magnetic) center.

6.9. After reassembly, perform final dielectric strength tests, megger, and note polarization index. Also, perform an ac high-potential test after the reconditioning is completed in order to detect defective insulation. The maximum allowable test voltage for a nondestructive high-potential test is 150% of rated line-to-line voltage.

6.10. Check dynamic balance and conduct a no-load test.

TECHNICAL REQUIREMENTS: REPLACEMENT AND REPAIRS

7.1. Items below are not covered under preventive maintenance but fall under the category of replacement and repairs. After determining those items that need replacement or repair and after notification to the purchaser in accordance with sections 5.5 and 5.6, the vendor shall perform the items below.

7.2. If the bearings require replacement, they shall be liberally sized and sealed to prevent entrance of dirt and the leakage of lubricant. Bearing measurement shall be recorded. Where necessary, the bearings shall be insulated to prevent circulation of shaft current through the bearings. All sleeve bearing housings shall have a drain plug, or equivalent, to facilitate flushing of the wells, to the extent that the original motors had them.

7.3. Vendor shall replace damaged thermocouple(s) on the motors which are so equipped with a chromel-constantan thermocouple to sense internal bearing metal temperature.

7.4. Vendor shall replace damaged bearing resistance-temperature detectors as required.

7.5. Vendor shall replace any power or instrument wiring that is degraded.

7.6. Vendor shall replace any strip heaters that are damaged or electrically inoperable.

TESTS, INSPECTIONS, AND DOCUMENTATION

8.1. The purchaser shall, at any reasonable time, be permitted to have his representative visit the vendor's factory and repair shop to ascertain if the material and processes used in its service conform to the specification and to discover whether if work is progressing at a proper rate to meet schedules.

8.2. The following tests and inspections, as a minimum, are to be performed by the vendor prior to shipment and the data recorded on Appendix 1, Quality Check Sheet.

1. Visual inspection
2. Megger test
3. High-potential test
4. Locked rotor current test
5. No-load running current test
6. Core and coil inspection
7. Dielectric strength test and polarization index
8. Air-gap readings
9. Phase-to-phase low-resistance check

8.3. The inspector shall use Appendix 1 (one for each motor) to assure that the indicated tests and procedures have been conducted and for a permanent record of what shop inspection procedures have been taken to provide quality equipment.

8.4. A complete inspection report with all test results and data sheets shall be provided for each motor. All documentation shall be sent to the purchaser at the location identified in the proposal to bid. The purchaser reserves the right to witness factory tests.

QUALITY CHECK SHEET

9.1. Appendix 1, Quality Check Sheet, is an integral part of this specification.

Appendix 1

Quality Check Sheet

Station ______ Date rcvd. ______ Date shipped ______
Cust. P.O. ______ Service ______ NYSS R.O. ______
Motor ______ hp ______ V, ______ Ph., ______ Cyc., ______ Ins. ______ Rise
Mfgr. ______ Model# ______ Ser. No. ______ Form ______
Horiz. ______ Vert. ______ Ball, ______ Sleeve brg. ______

Tests before Cleaning

Polarization index = 10 min ______ U_a = ______ PI, Megger ______
1 min ______ U_a
Air-gap readings PE ______ OPE ______

Disassembly Inspection

______ Stator cleaned with ______
______ Stator inspected for cleanliness
______ Stator iron, wedges, end bracing inspected
______ Stator varnish treated and baked
______ Leads examined, lead entrance checked for burrs
______ Rotor lathe check: Journal dia., PE ______ , OPE ______
______ Rotor bars inspected for cracks or breaks, iron checked for tightness
______ Rings turned polished (if applicable)
______ Rotor balanced (see balance report)
______ Bearings cleaned, examined, and measured (if sleeve brg.)
PE Hor. ______ , Vert. ______ ; OPE Hor. ______ , Vert. ______
______ Space heaters checked for condition of leads and operation
______ Hardware and fittings examined
______ Oil rings examined for concentricity
______ Coupling examined for rabbet fit and/or wear

Inspection after assembly

Polarization index = 10 min ______ U_a = ______ PI, Megger ______
1 min ______ U_a High-potential test completed ______

Line volts Ph1 ______ Ph2 ______ Ph3 ______
Line amps Ph1 ______ Ph2 ______ Ph3 ______
Vibration PE ______, OPE ______, No-load rpm ______
______ End play satisfactory (if sleeve bearing)
______ Oil rings turning (if sleeve bearing)
______ Lead identification legible and proper
______ Motor masked and painted properly
______ Oil drained and motor tagged (if oil lube)
______ Coupling installed properly (if shipped with motor)
______ Magnetic center mark verified or remarked
______ Space heaters checked for operation ______ amps
______ New bearing PE Hor. ______ Vert. ______ ; OPE Hor. ______ Vert. ______
______ Air-gap reading PE ______ OPE ______

Motor checked and ready for shipment

Index